Texts in Computing

Volume 21

A Mathematical Primer
on Computability

Texts in Computing Series Editor
Ian Mackie   mackie@lix.polytechnique

A Mathematical Primer
on Computability

Amílcar Sernadas
Cristina Sernadas
João Rasga
and
Jaime Ramos

ISBN 978-1-84890-296-1

College Publications
Scientific Director: Dov Gabbay
Managing Director: Jane Spurr

http://www.collegepublications.co.uk

Cover produced by Laraine Welch

Preface

The main objective of this book is to provide a self-contained introduction to computability theory to undergraduate students of Mathematics and Computer Science and is equally suitable for Science and Engineering students who are interested in gaining a deeper understanding of the mathematical aspects of the subject. The technical material is illustrated with plenty of examples, problems with fully-worked solutions as well as a range of proposed exercises. Applications of computability in several areas namely in logic, Euclidean geometry and graphs are presented. Nevertheless no previous knowledge of these subjects is required. The essential details for understanding the application are always provided.

We adopt an abstract high-level programming language for defining computable functions taking advantage that nowadays the students are familiar with basic primitives of high-level programming languages (such as simple loops and recursion). Moreover, important concepts can be exemplified in a very down to earth manner. Another advantage of this computational model is that we can work beyond the setting of natural numbers (although everything could be done in this setting).

The book is divided in two parts. Part I is centered around fundamental computability notions and results. Among the pillar concepts are computational model, computable function, decidable and listable set, proper universal function as well as decision problem. In Chapter 1 we discuss the important concept of computable function as well as operations on functions that preserve computability. Chapter 2 concentrates on decidable and listable sets and presents the important technique of problem reduction for transferring (positive and negative) decidability and listability results. A first example of problem reduction is provided for Gödelizations. Chapter 3 is targeted at the crucial notion of proper universal function. The concept is illustrated by univ which is a proper universal function defined in the adopted abstract high-level programming language. The first example of a listable non-decidable set is provided as well as an example of a non-listable set. The chapter ends with the introduction of proper universal set as well as preliminary relationship with proper universal functions. Chapter 4 is centered on some essential theorems. It starts with three main results on decidability and listability of sets of indices of computable functions over a proper universal function: Rice's Theorem (a necessary and sufficient condition for a set of indices to be decidable), Rice-Shapiro's Theorem (a necessary condition for a set of indices to be listable) and Rice-Shapiro-McNaughton-Myhill's Theorem (a necessary and sufficient condition for a set of indices to be listable). After that, Rogers' Theorem is established proving that any two proper universal functions are isomorphic. The chapter ends with two versions of the Recursion Theorem for proving sufficient conditions that guarantee the existence of some particular indices. Chapter 5 concentrates on many-to-one reducibility and many-to-one degrees. Furthermore it also includes notions like many-to-one complete sets, effectively non-listable sets as well as their relationship. The chapter goes on with the Post's construction for showing that there is a simple set. The chapter ends with productive sets and their relationship to effectively non-listable sets and complete sets. In Chapter 6 we discuss properties of operators that map unary functions to unary functions. The chapter starts with an introduction to the topic of computation with oracles. After that the important concept of computable operator is presented. Examples are provided of non-computable operators. Others kinds of operators are also discussed like monotonic and finitary operators. The relationship between the classes of operators is proved, including Myhill-Shepherdson's Theorem. The chapter ends with Myhill-Shepherdson's Existence Theorem as well as Kleene's Least Fixed Point Theorem. Finally, Part I terminates in Chapter 7 with a briefing on Turing computability. Namely, we discuss the notion of Turing computable function, Turing reducibility (as well as the comparison with many-to-one reducibility) and Turing degrees. We do not elaborate on Turing universal machines (the interested reader can see [44, 8, 53, 23]). All chapters of Part I terminate with a selected collection of solved problems as well some exercise proposals.

Part II of the book concentrates on illustrating the role of computability in other selected areas of Mathematics suitable for late undergraduate students. It is well known that computability has a strong relationship with logic in particular because of Gödel's Incompleteness Theorems ([12, 7, 19, 13, 14]). So, it comes as no surprise that we include logic herein. We start in Chapter 8 with first-order logic namely proving that the set of (Hilbert) consequences of a decidable set of formulas is listable. Then we prove Gödel's First Incompleteness Theorem as a relevant example of the importance of computability. Chapter 9 is targeted at establishing (positive or negative) decidability results of several decision problems starting with the simple case of prime numbers. Then, the well known satisfiability problem in propositional logic is discussed. We go on analyzing the consequence problem (with a non-empty set of hypotheses) in propositional logic. After that, we concentrate on the k-colouring of a finite graph as well as on the first-order theory of Euclidean geometry. In the last section we discuss the satisfiability problem for modal logic K. Part II terminates with a mathematical introduction to Kolmogorov Complexity in Chapter 10. By defining an appropriate family of universal functions over finite sets of oracles it is possible to get upper bounds for the complexity of programs for computable functions (relatively to finite sets of oracles).

Acknowledgements

We would like to express our deepest gratitude to the many undergraduate students of Instituto Superior Técnico that attended the Computability Theory course. We are also grateful to our colleague and friend Walter A. Carnielli for many discussions on computability theory in general and Gödel's Incompleteness Theorems in particular. We also acknowledge the interaction with André Souto on Kolmogorov complexity. Last but not least, we greatly acknowledge the excellent working environment provided by the Department of Mathematics of Instituto Superior Técnico.

The preparation of this book was overshadowed by Amilcar's death in February 2017. We had intended to write it jointly. Most of the ideas were worked out together and we had done our best to complete them. In sorrow we dedicate this book to his memory.

Lisbon,   *Cristina Sernadas*
October 2018   *João Rasga*
                                                     Jaime Ramos

Contents

Part I

Fundamentals of Computability

Chapter 1

Computable Functions

Intuitively a function is computable if there is a procedure that calculates its values. That is, when the function f is defined for a given argument w the execution of the procedure on w will necessarily terminate in a finite number of steps returning $f(w)$. Otherwise, if f is not defined for w then the execution of the procedure on w either does not terminate or it does but no meaningful result will be returned. So function f can be computable even when it is partial (possibly not defined everywhere).

For the moment, by a procedure we mean a program composed by a finite sequence of basic commands. An algorithm is a procedure whose execution always terminates in a finite number of steps, for the relevant input data.

For instance, there is an algorithm for determining whether or not a natural number is even based on the known fact that a natural number x is even if and only if the remainder of the division of x by 2 is zero. On the other hand, consider the function $f = \lambda x.x/2$ over the natural numbers (we use the lambda notation: $\lambda x.x/2$ is the function that maps x to $x/2$, see [4]). This function is defined whenever x is an even number and undefined otherwise. Thus, f is a partial function and is computable since there is a procedure such that its execution over an even number x returns $x/2$ and its execution elsewhere does not terminate.

For defining computable functions we need to adopt a *computational formalism* for describing procedures and algorithms. We adopt the Chuch-Turing Postulate (see [34]) stating that any computational formalism should be equivalent to Turing's computational model. However, some of them are more flexible than the others taking into account the functions we want to consider and the results we want to obtain.

In the examples above we are in the setting of natural numbers. This means that we concentrate on functions from $\mathbb{N}^n$ to $\mathbb{N}^m$ where $n \in \mathbb{N}$ and $m \in \mathbb{N}^+$. In this setting, Turing and Kleene models of computation are commonly used. Showing that a function is computable using Turing machines, see [63], consists of finding a Turing machine that computes the function. On the other hand, proving that a function is

3

computable using Kleene recursive functions, see [33], means that we must prove that the function is recursive which consists of writing a program where the construction of the function is reflected.

However, in many cases, we want to work outside the natural numbers setting (although everything we do could be done in this setting). For instance, we can be interested in reasoning about formulas and theorems in a logic from a decidable point of view. In this case, we have two possibilities: either we convert formulas and theorems to the setting of natural numbers via a Gödelization or we can define the concept of computable function in the appropriate setting by adopting a more flexible computational formalism. In both cases we must identify the alphabet at hand and define the set of formulas as a subset of the set of all finite sequences that we can write with the alphabet.

In this book we use an abstract high-level programming language for writing procedures and algorithms following closely mathematical notations. We omit non essential details and simplify the presentation so that the book can be read by anyone with a very shallow contact with programming.

For showing that a function is computable, no matter the setting, we must find a procedure that computes the function.

1.1 Preliminaries

Given a set S and $s_1, \ldots, s_n \in S$, we denote by $|s_1 \ldots s_n|$ the *length* n of the sequence $s_1 \ldots s_n$ over S. Moreover, we denote by ε the *empty sequence*, that is, the sequence with length 0. Finally, we denote by

$$S^*$$

the *set of all finite sequences* over S also called the *universe* over S.

We assume fixed a finite non empty alphabet

$$A$$

containing capital and small letters, digits, punctuation marks, whitespace characters and logical, numerical and comparison symbols. We denote by

$$W$$

the universe A^*.

Example 1.1. For instance, the sequences

$$\mathsf{function}(x)(\mathsf{return}\,1) \quad \text{and} \quad \mathsf{functionwhile}$$

are in W.

In order to improve readability, letters in A may be presented in sans-serif font as we did in the example above.

Definition 1.1. A *working universe* is an infinite subset of W.

Example 1.2. For instance $\mathbb{N}, \mathbb{Z}, \mathbb{Q}, \{1\}^*, \{0,1\}^*$ are working universes.

Besides the expected working universes in the previous example we now introduce a completely different working universe.

Example 1.3. The *language L_P of propositional logic* ([15, 55]) over a set $P = \{p_0, p_1, \dots\}$ of propositional symbols is inductively defined as follows:

- $P \subseteq L_P$;

- $(\neg \alpha) \in L_P$ whenever $\alpha \in L_P$;

- $(\alpha_1 \supset \alpha_2) \in L_P$ whenever $\alpha_1, \alpha_2 \in L_P$.

The working universe W_{L_P} for propositional logic is inductively defined as follows:

- $\langle "p", k \rangle \in W_{L_P}$ whenever $p_k \in P$;

- $\langle "not", \alpha \rangle \in W_{L_P}$ whenever $\alpha \in W_{L_P}$;

- $\langle "implies", \alpha_1, \alpha_2 \rangle \in W_{L_P}$ whenever $\alpha_1, \alpha_2 \in W_{L_P}$.

For instance,

$$\langle "implies", \langle "p", 1 \rangle, \langle "p", 2 \rangle \rangle \in W_{L_P}$$

is the representation of the formula $(p_1 \supset p_2)$ in the working universe W_{L_P}.

In order to simplify the presentation, we may refer to elements in W_{L_P} as the corresponding ones in L_P. Thus, from now on, by L_P we mean W_{L_P}. We may also apply this simplification to other working universes.

Definition 1.2. Let W be a working universe and $n \in \mathbb{N}^+$. A set $C \subseteq W^n$ is called a *set of type n over W* or simply a *set over W* if no confusion arises.

Definition 1.3. Let W_1 and W_2 be working universes, C_1 and C_2 sets over W_1 and W_2, respectively, and f a function from C_1 to C_2. The *domain* of f is the set

$$\mathrm{dom}\, f = \{w \in C_1 : f(w) \downarrow \text{ and } f(w) \in C_2\}$$

where the notation $f(w) \downarrow$ indicates that f is defined for w. The *range* of f is

$$\mathrm{range}\, f = f(\mathrm{dom}\, f).$$

Moreover, we say that C_1 and C_2 are the *source* and the *target* sets of f, respectively.

Remark 1.1. We use the notation

$$f : C_1 \rightharpoonup C_2$$

to indicate that f is a function from C_1 to C_2 (that can be defined on a proper subset of C_1). We say that f is a *map or a total function*, written

$$f : C_1 \rightarrow C_2$$

whenever dom $f = C_1$.

Remark 1.2. We use the lambda notation for defining functions. For instance,

$$\lambda x, y . x + y : \mathbb{N}^2 \rightarrow \mathbb{N}$$

is the function

$$(x, y) \mapsto x + y : \mathbb{N}^2 \rightarrow \mathbb{N}$$

that associates to each pair of natural numbers their sum. The lambda notation allows the presentation of all the details of a function in a concise way.

Example 1.4. Let E be the set of all even natural numbers. The following are illustrations of functions:

$$f = \lambda x . \frac{x}{2} : \mathbb{N} \rightharpoonup \mathbb{N}$$

where dom $f = E$ and range $f = \mathbb{N}$ and

$$g = \lambda x . \begin{cases} \frac{x}{2} & \text{if } x \in (E \setminus \{4\}) \\ \text{undefined} & \text{otherwise} \end{cases} : \mathbb{N} \rightharpoonup \mathbb{N}$$

where dom $g = E \setminus \{4\}$ and range $g = \mathbb{N} \setminus \{2\}$.

Example 1.5. Let

$$f = \lambda x . \sqrt{x} : \mathbb{N} \rightharpoonup \mathbb{N}.$$

Then dom $f = \{x^2 : x \in \mathbb{N}\}$ and range $f = \mathbb{N}$. On the other hand, consider

$$g = \lambda x . \begin{cases} 5 & \text{if } x \neq 10 \\ -1 & \text{otherwise} \end{cases} : \mathbb{N} \rightharpoonup \mathbb{N}.$$

Hence dom $g = \mathbb{N} \setminus \{10\}$ and range $g = \{5\}$.

Definition 1.4. The function

$$\lambda w . \text{undefined} : C_1 \rightharpoonup C_2$$

is called the *undefined function* from C_1 to C_2.

Observe that dom λw. undefined $=$ range λw. undefined $= \emptyset$. Moreover, λw. undefined is the unique function with empty domain and range.

Definition 1.5. Let W be a working universe and $C \subseteq W^n$. Then, the functions

$$\chi_{C:W} = \lambda w. \begin{cases} 1 & \text{if } w \in C \\ 0 & \text{otherwise} \end{cases} : W^n \to \mathbb{N}$$

and

$$\chi^p_{C:W} = \lambda w. \begin{cases} 1 & \text{if } w \in C \\ \text{undefined} & \text{otherwise} \end{cases} : W^n \rightharpoonup \mathbb{N}$$

are called the *characteristic map* and the *characteristic function* of C, respectively.

Observe that
$$\text{dom}\,\chi_{C:W} = W^n \text{ and range}\,\chi_{C:W} \subseteq \{0,1\}$$
and
$$\text{dom}\,\chi^p_{C:W} = C \text{ and range}\,\chi^p_{C:W} \subseteq \{1\}.$$

Thus, $\chi_{C:W}$ and $\chi^p_{C:W}$ are completely different functions. When W is $\mathbb{W}$ we will write

$$\chi_C \text{ and } \chi^p_C$$

instead of $\chi_{C:\mathbb{W}}$ and $\chi^p_{C:\mathbb{W}}$, respectively. Moreover, we also need the maps J and zigzag introduced below.

J	0	1	2	3	...
0	0	1	3	6	...
1	2	4	7	11	...
2	5	8	12	17	...
3	9	13	18	24	...
...	...	...	...	...	...

Figure 1.1: Bijection J between $\mathbb{N}^2$ and $\mathbb{N}$

Remark 1.3. We denote by
$$J$$
the map $\lambda i, j. i + \frac{1}{2}((i+j)(i+j+1)) : \mathbb{N} \times \mathbb{N} \to \mathbb{N}$ and by
$$\text{zigzag} : \mathbb{N} \to \mathbb{N}^2$$

its inverse. Moreover, we denote by

$$K, L : \mathbb{N} \to \mathbb{N}$$

the first and the second projection of zigzag.

In Figure 1.1, we illustrate the result of J on some values.

Exercise 1.1. Show that $J : \mathbb{N}^2 \to \mathbb{N}$ is a bijection.

Finally, we introduce the notion of subfunction/extension of partial functions.

Definition 1.6. Given $f_1, f_2 \in \mathscr{F}_1$, we say that f_1 is a *subfunction* of f_2 or that f_2 is an *extension* of f_1, written

$$f_1 \subseteq f_2,$$

whenever $f_2(x) = f_1(x)$ for every $x \in \mathrm{dom}\, f_1$. We say that f is *equal* to g, written $f = g$ whenever $f \subseteq g$ and $g \subseteq f$.

Clearly, when $f \subseteq g$ then $\mathrm{dom}\, f_1 \subseteq \mathrm{dom}\, f_2$. On the other hand, when $f = g$ then $\mathrm{dom}\, f = \mathrm{dom}\, g$. Moreover, when $f = g$ then $f(x) = g(x)$ for every $x \in \mathrm{dom}\, f$ but no condition is imposed on $f(x)$ and $g(x)$ when $x \notin \mathrm{dom}\, f$ except that neither of them should be in $\mathbb{N}$.

Example 1.6. Given $C \subseteq W^n$, we have that $\chi^{\mathrm{p}}_{C:W} \subseteq \chi_{C:W}$.

1.2   Computational Model

We adopt as model of computation an abstract high-level programming language over the alphabet A. A *program* or a *procedure* is an element in W of the form

$$\mathrm{function}\,(args)\,(body)$$

where *args* is a finite list of variables and *body* is a non-empty finite sequence of commands (separated by ;).

We use five main commands: return statement, assignment, loop, alternative and null statement. The return statement command

$$\mathrm{return}\ exp_1, \ldots, exp_n$$

when executed, stops the execution of the function and returns the values of expressions $exp_1, \ldots, exp_n$.

Example 1.7. Consider the following program P_+

$$\mathrm{function}\ (x, y)\ (\mathrm{return}\ 2x + 3y)$$

The execution of P_+ on (k_1, k_2) terminates returning the value $2k_1 + 3k_2$.

Example 1.8. Consider the following program

$$\text{function } (s_1, s_2) \ (\text{return } s_1 \cdot "+" \cdot s_2)$$

where the operation $\cdot$ when receiving two strings returns the concatenation of the strings. The execution of P on strings s_1 and s_2 returns the string composed by the symbols in s_1 followed by the symbol $+$ followed by the symbols in s_2.

The assignment command

$$var_1, \ldots, var_n = exp_1, \ldots, exp_n$$

when executed, assigns simultaneously the values of the expressions $exp_1, \ldots, exp_n$ to the variables $var_1, \ldots, var_n$.

Example 1.9. Consider the following program P

$$\text{function } (n) \ ($$
$$f = \text{function } (x, y) \ (\text{return } 2x + 3y);$$
$$\text{return } n + f(n, n+1)$$
$$)$$

The execution of P on a natural number n returns $6n + 3$. An alternative program doing the same thing as P is as follows

$$\text{function } (n) \ (\text{return } n + P_+(n, n+1))$$

where P_+ is defined in Example 1.7.

The loop command

$$\text{while } cond \text{ do } (body)$$

when executed, repeatedly executes the commands in the *body* while the Boolean expression *cond* is true. To simplify the presentation we may omit the parentheses when the body has just one command.

Example 1.10. Consider the following program P_{fact}

```
function (n) (
    k = 1;
    r = 1;
    while k ≤ n do (
        r = r × k;
        k = k + 1
    );
    return r
)
```

The execution of P on a natural number n returns $n!$. An alternative solution could be achieved by using recursion:

$$P_{factr} = \text{function } (n) \ (\text{if } n == 0 \text{ then return } 1 \text{ else return } n \times P_{factr}(n-1))$$

The null statement

$$\text{null}$$

when executed, does nothing.

Example 1.11. Consider the following program

$$\text{function } (n) \ (\text{while true do null})$$

The execution of this program does not terminate since the condition of the loop is always true.

The alternative command

$$\text{if } cond \text{ then } (body_1) \text{ else } (body_2)$$

when executed, executes the commands in $body_1$ when the Boolean expression $cond$ is true and executes the commands in $body_2$ otherwise. To simplify the presentation we may omit the parentheses when $body_2$ has just one command. A more simplified version of the alternative command is

$$\text{if } cond \text{ then } (body)$$

used if no action is to be taken when $cond$ is false. To simplify the presentation we may omit the parentheses when there is no ambiguity and $body$ has just one command.

Example 1.12. Consider the following program P

```
function (w) (
    if w == ⟨⟩ then
        return ⟨⟩;
    if w[1] % 2 == 0 then
        return ⟨w[1]/2⟩;
    return ⟨(w[1] − 1)/2⟩
)
```

where the operation % when applied to natural numbers m and n returns the remainder of the integer division of m by n and $==$ is the equality verification operator. The execution of P on a list w returns the empty list $\langle \rangle$ if w is empty and returns the list with the first element $w[1]$ divided by 2 when $w[1]$ is even and $w[1] - 1$ divided by 2 otherwise.

In what follows we identify an expression with its evaluation. For instance, $2+3$ is identified with 5. Furthermore, when P is the program function (k) (return $k+1$) then $\mathsf{P}(3)$ is identified with 4. Observe that, for instance, the evaluation of not(α) is not(α) since, although syntactically correct, no expression is assigned to not.

We assume that the abstract high-level programming language contains the usual predefined programs for numerical, logical, string and list manipulation, namely:

- isnat, that when executed on e, returns true when e is a natural number and returns false otherwise;

- stceval$_n^m$ for $m, n \in \{1, 2\}$, the *success and time constrained n-ary evaluation by m programs*, that when executed on programs $\mathsf{P}_1, \ldots, \mathsf{P}_m$, expressions $e_1, \ldots, e_n$, a natural number t and an expression v, returns:

  - the value of v if the execution of $\mathsf{P}_m(\ldots \mathsf{P}_1(e_1, \ldots, e_n) \ldots)$ does not terminate within t units of time;
  - the result of the execution of $\mathsf{P}_m(\ldots \mathsf{P}_1(e_1, \ldots, e_n) \ldots)$, otherwise.

  In order to simplify the presentation we omit the reference to n and m when there is no ambiguity;

- pos, that when executed on a list w and on an expression e, returns the position of e in w providing that e occurs in w and, otherwise, returns 0;

- append, that when executed on a list w and an expression e, returns the list with the elements of w followed by e;

- length, that when executed on a list w, returns the number of elements of w;

- subs, that when executed on expressions e_1, e_2, e_3, returns the expression obtained by replacing e_2 by e_3 in e_1.

We now introduce a program that illustrates the use of predefined programs on lists.

Example 1.13. Consider the following program $\mathsf{P}_{\text{count}}$

```
function (w, e) (
    i = 1;
    r = 0;
    while i ≤ length(w) do (
        if w[i] == e then
            r = r + 1;
        i = i + 1
    );
    return r
)
```

The execution of P_{count} on a list w and an expression e returns the number of occurrences of e in w.

1.3 Key Concept

We are ready to introduce the important concept of computable function.

Definition 1.7. Let C_1 and C_2 be sets of types m and n, respectively. A function $f : C_1 \rightharpoonup C_2$ is said to be *computable* if there is a procedure P written in the programming language that given an element $(w_1, \ldots, w_m)$ in C_1

- if $(w_1, \ldots, w_m) \in \mathrm{dom}\, f$ and $f(w_1, \ldots, w_m) = (v_1, \ldots, v_n)$ then the execution of P on $w_1, \ldots, w_m$ terminates in a finite number of steps returning $v_1, \ldots, v_n$ as the result;

- if $(w_1, \ldots, w_m) \notin \mathrm{dom}\, f$ then the execution of P on $w_1, \ldots, w_m$:

    - either does not terminate;
    - or terminates in a finite number of steps with a result not in C_2.

In this case, we say that function f is *computed* by P. We may also say that P *computes* f or P *is a program for* f. It should be stressed that nothing is imposed on the behavior of P when given an input not in C_1 besides stating that the result (if it exists) is not in C_2. Observe that a computable function f can be computed by several different procedures.

Definition 1.8. A procedure P that computes a map f is said to be an *algorithm* for f.

Example 1.14. Let E be the set of even natural numbers. Consider the characteristic map $\chi_{E:\mathbb{N}}$ (recall Definition 1.5). The program P

$$\text{function } (x) \ (\text{if } x\%2 == 0 \text{ then return } 1 \text{ else return } 0)$$

computes $\chi_{E:\mathbb{N}}$ as we show now. Let $n \in \mathbb{N}$.

(1) Assume that n is an even number. Hence, when executing P on n, the condition of the if is true and so P returns 1 which is $\chi_{E:\mathbb{N}}(n)$.

(2) Assume that n is an odd number. Hence, when executing P on n, the condition of the if is false and so P returns 0 which is $\chi_{E:\mathbb{N}}(n)$.

Thus $\chi_{E:\mathbb{N}}$ is a computable map.

Observe that in this case the execution of P on something other than a natural number terminates on a finite number of steps returning 0. That is, P returns the same value 0 whenever the argument is either an odd number or not a natural number. In alternative we can consider the following program P$'$

```
function (x) (
    if isnat(x) then
        if x % 2 == 0 then return 1 else return 0
    else
        while true do null
)
```

The execution of P′ on x terminates in a finite number of steps with 1 whenever x is an even natural number and with 0 whenever x is an odd natural number. Otherwise, the execution of P′ does not terminate.

Example 1.15. Consider the undefined function in Definition 1.4. The following program P

$$\text{function } (w) \text{ (while true do null)}$$

computes $\lambda w . \text{undefined} : C_1 \rightarrow C_2$. Indeed, let w be an arbitrary element of C_1. Then $w \notin \text{dom} \, \lambda w . \text{undefined}$. Note that the execution of P on w does not terminate. Therefore, P computes $\lambda w . \text{undefined}$ and so $\lambda w . \text{undefined}$ is a computable function.

Example 1.16. Consider function $f : \mathbb{N} \rightharpoonup \mathbb{N}$ as defined in Example 1.4. The following program P

$$\text{function } (x) \text{ (if } x \% 2 == 0 \text{ then return } x/2 \text{ else while true do null)}$$

computes f. Indeed, let $n \in \mathbb{N}$.

(1) Assume that $n \in \text{dom} \, f$. Then, by definition of f, n is an even number. Hence, when executing P on n, the guard of the if is true and P returns $\frac{n}{2}$ which is $f(n)$.

(2) Assume that $n \notin \text{dom} \, f$. Then, by definition of f, n is an odd number. Hence, when executing P on n, the guard of the if is false and P will run forever since the loop while true do null does not terminate.

Hence, f is a computable function.

As an alternative let P′ be the program defined as follows:

$$\text{function } (x) \text{ (if } x \% 2 == 0 \text{ then return } x/2 \text{ else return } -1)$$

Observe that, when x is an odd number, the execution of P′ on x returns -1 which is outside the target set $\mathbb{N}$. Otherwise P′ returns $\frac{x}{2}$. So P′ also computes f.

Exercise 1.2. Recall the maps J, zigzag, K and L introduced in Notation 1.3. Show that these maps are computable.

Exercise 1.3. Show that $\chi_{\mathbb{N}^n}$ is a computable map for each $n \in \mathbb{N}^+$.

1.4   Operations on Computable Functions

We now investigate some properties of the class of computable functions, namely operations between computable functions that return computable functions. We start by introducing the operation restriction of a function.

Definition 1.9. Let $f : C_1 \rightharpoonup C_2$ be a function and $S \subseteq C_1$ a set. The *restriction* of f to S is the function

$$f|_S : S \rightharpoonup C_2$$

such that

- $\operatorname{dom} f|_S = \operatorname{dom} f \cap S$;

- $f|_S(w) = f(w)$ whenever $w \in \operatorname{dom} f|_S$.

Example 1.17. Consider function $f : \mathbb{N} \rightharpoonup \mathbb{N}$ defined in Example 1.4 and program P in Example 1.16 that computes f. Let

$$f|_E : E \to \mathbb{N}$$

be the restriction of f to the set E of even numbers. We show that procedure P computes $f|_E$. Indeed, let $x \in E$. Then, $x \in \operatorname{dom} f|_E = \operatorname{dom} f = E$. So the execution of P on x will always terminate in a finite number of steps with $f|_E(x)$ since $f|_E(x) = f(x)$. Observe that P is an algorithm for $f|_E$. Therefore, we can conclude that $f|_E$ is a computable map.

Proposition 1.1. Let $f : C_1 \rightharpoonup C_2$ be a function and $S \subseteq C_1$. Assume that f is a computable function. Then, $f|_S$ is a computable function.

Proof. Suppose that $f : C_1 \rightharpoonup C_2$ is a computable function and $S \subseteq C_1$. Let P_f be a program that computes f. We show that P_f computes $f|_S : S \rightharpoonup C_2$. Let $a \in S$.

(1) Assume that $a \in \operatorname{dom} f|_S$. Then, $a \in S \cap \operatorname{dom} f$ and $f|_S(a) = f(a)$. Thus, the execution of P_f on a terminates within a finite number of steps returning $f(a) = f|_S(a)$.

(2) Assume that $a \notin \operatorname{dom} f|_S$. Then, $a \notin S \cap \operatorname{dom} f$. Hence, $a \notin \operatorname{dom} f$. Then the execution of P_f either does not terminate or it terminates in a finite number of steps with a result not in C_2. $\qquad\square$

Remark 1.4. Given a map $h : S_1 \to S_2$, we denote by

$$h^\bullet$$

the map from S_1 to $(\operatorname{range} h)$ such that $h^\bullet(s) = h(s)$.

Exercise 1.4. Show that $h^\bullet$ is computable whenever h is computable.

Composition

We now introduce and illustrate composition of functions.

Definition 1.10. Let $f : C_1 \rightharpoonup C_2$ and $g : C_2 \rightharpoonup C_3$ be functions. Then, the *composition* of f and g is the function $g \circ f : C_1 \rightharpoonup C_3$ such that

- $\operatorname{dom} g \circ f = \{w \in C_1 : w \in \operatorname{dom} f \text{ and } f(w) \in \operatorname{dom} g\}$;
- $(g \circ f)(w) = g(f(w))$, whenever $w \in \operatorname{dom} g \circ f$.

Example 1.18. Consider the functions

$$f = \lambda x. \begin{cases} 5 & \text{if } x \geq 5 \\ 4 & \text{if } x = 3 \\ \text{undefined} & \text{otherwise} \end{cases} , \quad g = \lambda x. \begin{cases} 4 & \text{if } 5 \leq x < 7 \\ -1 & \text{if } x = 4 \\ \text{undefined} & \text{otherwise} \end{cases} : \mathbb{N} \rightharpoonup \mathbb{N}.$$

Then, the composition of f and g is

$$g \circ f = \lambda x. \begin{cases} 4 & \text{if } x \geq 5 \\ \text{undefined} & \text{otherwise} \end{cases} : \mathbb{N} \rightharpoonup \mathbb{N}$$

since

- $\operatorname{dom} g \circ f = \{x \in \mathbb{N} : x \geq 5\}$;
- $g \circ f(x) = \{4\}$ for every $x \in \operatorname{dom} g \circ f$.

Proposition 1.2. Assume that $f : C_1 \rightharpoonup C_2$ and $g : C_2 \rightharpoonup C_3$ are computable functions and that χ_{C_2} is a computable map. Then, $g \circ f$ is also a computable function.

Proof. Assume that P_f, P_g and $P_{\chi_{C_2}}$ compute f, g and χ_{C_2}, respectively. Consider the following program $P_{g \circ f}$:

function (w) (if $P_{\chi_{C_2}} (P_f(w)) == 1$ then return $P_g(P_f(w))$ else while true do null)

We show that $P_{g \circ f}$ computes $g \circ f$. Let $w \in C_1$.
(1) Assume that $x \in \operatorname{dom} g \circ f$. Then, $w \in \operatorname{dom} f$ and $f(w) \in \operatorname{dom} g$. Thus, the execution of P_f on w terminates in a finite number of steps returning $f(w) \in C_2$. Hence, when executing $P_{g \circ f}$ on w, the guard of the if is true. Moreover, the execution of P_g on $f(w)$ terminates in a finite number of steps returning $g(f(w))$.

(2) Assume that $w \notin \operatorname{dom} g \circ f$. Then there are two cases to consider:
(a) $w \notin \operatorname{dom} f$. If the execution of P_f on w does not terminate then the execution of $P_{g \circ f}$ on w does not terminate as well. Otherwise, if the execution of P_f on w terminates in a finite number of steps with a result not in C_2 then, when executing $P_{g \circ f}$ on w, the guard of the if is false and so the execution of $P_{g \circ f}$ does not terminate.

(b) $f(w) \notin \operatorname{dom} g$ and $w \in \operatorname{dom} f$. Then, the execution of P_g on $f(w)$ either does not terminate or it terminates in a finite number of steps with a result not in C_3. $\qquad \square$

Composition can be extended to finite sets of functions even in the presence of partiality. It is enough to prove that composition is associative.

Proposition 1.3. Let $f : C_1 \rightharpoonup C_2$, $g : C_2 \rightharpoonup C_3$ and $h : C_3 \rightharpoonup C_4$ be arbitrary functions. Then,

$$h \circ (g \circ f) = (h \circ g) \circ f.$$

Proof. We must prove two facts:

(1) $\operatorname{dom} h \circ (g \circ f) = \operatorname{dom}(h \circ g) \circ f$. Indeed,

$$
\begin{aligned}
\operatorname{dom} h \circ (g \circ f) & \\
= {} & \{w \in C_1 : w \in \operatorname{dom}(g \circ f), (g \circ f)(w) \in \operatorname{dom} h\} \\
= {} & \{w \in C_1 : w \in \operatorname{dom}(g \circ f), g(f(w)) \in \operatorname{dom} h\} \\
= {} & \{w \in C_1 : w \in \operatorname{dom} f, f(w) \in \operatorname{dom} g, g(f(w)) \in \operatorname{dom} h\} \\
= {} & \{w \in C_1 : w \in \operatorname{dom} f, f(w) \in \operatorname{dom} h \circ g\} \\
= {} & \operatorname{dom}(h \circ g) \circ f.
\end{aligned}
$$

(2) $h \circ (g \circ f)(w) = (h \circ g) \circ f(w)$. Follows immediately by the second condition on the definition of composition. □

So, from now on, given the set $\{f_1 : C_1 \rightharpoonup C_2, \dots f_n : C_n \rightharpoonup C_{n+1}\}$ of functions, we can refer to

$$f_n \circ \dots \circ f_1$$

as the composition of the functions in the set.

Exercise 1.5. Establish a sufficient condition for the computability of $f_n \circ \dots \circ f_1$.

Aggregation

We now discuss the preservation of computability by aggregation.

Definition 1.11. Let $f : C_1 \rightharpoonup C_2$ and $g : C_1 \rightharpoonup C_3$ be functions. Then, the *aggregation* of f and g is the function $\langle f, g \rangle : C_1 \rightharpoonup C_2 \times C_3$ such that

- $\operatorname{dom}\langle f, g \rangle = \operatorname{dom} f \cap \operatorname{dom} g$;

- $\operatorname{range}\langle f, g \rangle = \{(f(w), g(w)) : w \in \operatorname{dom}\langle f, g \rangle\}$.

Example 1.19. Recall Example 1.18. Then, the aggregation of f and g is

$$\langle f, g \rangle = \lambda x. \begin{cases} (5,4) & \text{if } 5 \leq x < 7 \\ \text{undefined} & \text{otherwise} \end{cases} : \mathbb{N} \rightharpoonup \mathbb{N}^2$$

since

- $\text{dom}\langle f, g \rangle = \{5, 6\}$;

- $\langle f, g \rangle(x) = (5, 4)$ for every $x \in \text{dom}\langle f, g \rangle$.

Proposition 1.4. Assume that $f : C_1 \rightharpoonup C_2$ and $g : C_1 \rightharpoonup C_3$ are computable functions. Then, $\langle f, g \rangle$ is also a computable function.

Proof. Assume that P_f and P_g compute f and g, respectively. Consider the following program $\mathsf{P}_{\langle f, g \rangle}$:

$$\text{function } (w) \ (\text{return } \mathsf{P}_f(w), \mathsf{P}_g(w))$$

We show that $\mathsf{P}_{\langle f, g \rangle}$ computes $\langle f, g \rangle$.
(1) Assume that $w \in \text{dom}\langle f, g \rangle$. Then, $w \in \text{dom} f$ and $w \in \text{dom} g$. Hence, the execution of P_f and P_g on w terminate in a finite number of steps returning $f(w)$ and $g(w)$, respectively. Thus, the execution of $\mathsf{P}_{\langle f, g \rangle}$ on w terminates in a finite number of steps returning $f(w), g(w)$.
(2) Assume that $w \notin \text{dom}\langle f, g \rangle$. Then either $w \notin \text{dom} f$ or $w \notin \text{dom} g$. There are two cases:
(a) $w \notin \text{dom} f$. Then, the execution of P_f on w either does not terminate or terminates in a finite number of steps with a result not in C_2. So the execution of $\mathsf{P}_{\langle f, g \rangle}$ on w either does not terminate or terminates in a finite number of steps with a result where the first component is not in C_2.
(b) $w \notin \text{dom} g$. This case is simlar to (a). □

Aggregation can easily be extended to a finite set of functions with the same source set.

Exercise 1.6. Assume $f : C_1 \rightharpoonup C_2$ and $g_i : C_2 \rightharpoonup C_i'$ are computable functions for $i = 1, \ldots, k$. Find a sufficient condition for

$$\langle g_1, \ldots, g_k \rangle \circ f : C_1 \rightharpoonup C_1' \times \cdots \times C_k'$$

to be a computable function.

Projection

In order to introduce the operations of projection and minimization, we need some additional notation.

Remark 1.5. Given $n, i, j \in \mathbb{N}^+$ with $i \leq j \leq n$ and a set $C \subseteq W^n$, we denote by

$$C[i, j] \subseteq W^{j - i + 1}$$

the set

$$\{(u_i, \ldots, u_j) : \exists u_1, \ldots, u_{i-1}, u_{j+1}, \ldots, u_n \ (u_1, \ldots, u_n) \in C\}.$$

Furthermore, for each $i = 1, \ldots, n$, we may write

$$C[i]$$

for the set $C[i,i] \subseteq W$.

Definition 1.12. Let $C \subseteq W^n$ be a non-empty set with $n \in \mathbb{N}^+$. A *projection* over C and $i, j \in \mathbb{N}^+$ with $i \leq j \leq n$ is the following map:

$$\text{proj}^C_{[i,j]} = \lambda\, u_1, \ldots, u_n \cdot (u_i, \ldots, u_j) : C \to C[i,j].$$

Proposition 1.5. The class of computable functions is closed under projections.

Proof. Assume that $C \subseteq W^n$ is a non-empty set with $n \in \mathbb{N}^+$. For each $i, j \in \mathbb{N}^+$ with $i \leq j \leq n$, the following algorithm:

$$\text{function } (w_1, \ldots, w_n)\ (\text{return } w_i, \ldots, w_j)$$

computes the map $\text{proj}^C_{[i,j]}$. $\square$

Example 1.20. For instance the *identity map*

$$id_C : C \to C$$

over $C \subseteq W^n$ is computable since it is $\text{proj}^C_{[1,n]}$.

Minimization

Before defining minimization we need to introduce some auxiliary notions.

Definition 1.13. Let $\preceq$ be a binary relation over set S. We say that $\preceq$ is *total* whenever it is

- *antisymmetric*, that is, for every $s_1, s_2 \in S$ if $s_1 \preceq s_2$ and $s_2 \preceq s_1$ then $s_1 = s_2$;

- *transitive*, that is, for every $s_1, s_2, s_3 \in S$ if $s_1 \preceq s_2$ and $s_2 \preceq s_3$ then $s_1 \preceq s_3$;

- *connex*, that is, for every $s_1, s_2 \in S$ either $s_1 \preceq s_2$ or $s_2 \preceq s_1$.

Moreover, $\preceq$ is a *well-founded relation* if every non-empty set $R \subseteq S$ has a minimal element.

The relation $\preceq$ induces a relation $\prec$ such that $s_1 \prec s_2$ iff $s_1 \preceq s_2$ and $s_1 \neq s_2$. Observe that S has a minimal element with respect to $\preceq$. Moreover, when S is an infinite set then for each element s of S there is always a unique element s' such that $s \prec s'$ and for any s'' with $s \prec s''$ then $s' \prec s''$. In the sequel we say that s' is *the least element* greater than s for $\preceq$.

Remark 1.6. Given an infinite set S and a well-founded total order $\preceq$ on S, we denote by

$$\lambda k . k_{(S, \preceq)}$$

the map from $\mathbb{N}$ to S such that

- $0_{(S, \preceq)}$ is the minimal element of S;

- $(k+1)_{(S, \preceq)}$ is the least element greater than $k_{(S, \preceq)}$ for $\preceq$.

Definition 1.14. Let $m \in \mathbb{N}^+$, $C_1 \subseteq W_1^{m+1}$ be a non-empty set, $\preceq$ a well-founded total order in $C_1[m+1]$, $C_2 \subseteq W_2^n$ a non-empty set, $w \in C_2$ and $f : C_1 \rightharpoonup C_2$. The *minimization* of f on w for $\preceq$ is the function

$$\min_{fw}^{\preceq} : C_1[1,m] \rightharpoonup C_1[m+1],$$

defined as follows:[1]

- $\min_{fw}^{\preceq}(v) = x$ if $f(v,x) = w$ and, for every $x' \prec x$, $(v,x') \in \mathrm{dom}\, f$ and $f(v,x') \neq w$;

- $\min_{fw}^{\preceq}(v)$ is otherwise undefined.

When $m = 0$, the *minimization* of f on w for $\preceq$ is the function

$$\min_{fw}^{\preceq} : \{\emptyset\} \rightharpoonup C_1,$$

defined as follows:

- $\min_{fw}^{\preceq}(\emptyset) = x$ if $f(x) = w$ and, for every $x' \prec x$, $x' \in \mathrm{dom}\, f$ and $f(x') \neq w$;

- $\min_{fw}^{\preceq}(\emptyset)$ is otherwise undefined.

Observe that, when $m \neq 0$, $\min_{fw}^{\preceq}$, for every v, returns the minimum x such that $f(v,x) = w$ when there is such an x, and is undefined otherwise.

Remark 1.7. When $\preceq$ is clear from the context, function $\min_{fw}^{\preceq}$ can be denoted by

$$\lambda v . (\mu x . f(v,x) = w)$$

when $m \neq 0$, and by

$$(\mu x . f(x) = w)$$

when $m = 0$.

Even if f is a map, it may be the case that $\min_{fw}^{\preceq}$ is a partial function as we now illustrate.

---

[1]Writing, as usual, (v,x) for $(v_1, \ldots, v_m, x)$ when v is an element of a set of type m.

Example 1.21. Let E be the set of even natural numbers,

$$f = \lambda x_1, x_2 . x_1 \times x_2 : E^2 \to E$$

and $\leq_E$ the restriction of the usual well-founded total order over $\mathbb{N}$ to E. Then,

$$\min_{f4}^{\leq_E} = \lambda x_1 . \begin{cases} 2 & \text{if } x_1 = 2 \\ \text{undefined} & \text{otherwise} \end{cases} : E \to E.$$

Example 1.22. Consider the function

$$f = \lambda x_1, x_2 . |x_1 - 2x_2| : \mathbb{N}^2 \to \mathbb{N}.$$

Let $\leq$ be the usual total and well-founded relation over $\mathbb{N}$. Then,

$$\min_{f0}^{\leq} = \lambda x_1 . (\mu x . f(x_1, x) = 0) = \lambda x_1 . \begin{cases} \text{undefined} & \text{if } x_1 \text{ is odd} \\ \dfrac{x_1}{2} & \text{otherwise} \end{cases} : \mathbb{N}^2[1] \to \mathbb{N}^2[2].$$

Definition 1.15. Given $\preceq$ and $f : C_1 \to C_2$ as in Definition 1.14, function

$$\min_f^{\preceq} = \begin{cases} \lambda v, w . \min_{fw}^{\preceq}(v) : C_1[1, m] \times C_2 \to C_1[m+1] & \text{if } m \neq 0 \\ \lambda w . \min_{fw}^{\preceq}(\emptyset) : C_2 \to C_1 & \text{otherwise} \end{cases}$$

is said to be the *parameterized minimization* of f for $\preceq$.

Remark 1.8. When $\preceq$ is clear from the context, function $\min_f^{\preceq}$ can be denoted by

$$\lambda v, w . (\mu x . f(v, x) = w)$$

when $m \neq 0$, and by

$$\lambda w . (\mu x . f(x) = w)$$

when $m = 0$.

Example 1.23. Consider function f as defined in Example 1.22. Then,

$$\min_f^{\leq} = \lambda x_1, w . (\mu x . |x_1 - 2x| = w)$$

$$= \lambda x_1, w . \begin{cases} \dfrac{x_1 - w}{2} & \text{if } |x_1 - w| \in E \text{ and } x_1 \geq w \\ \dfrac{w - x_1}{2} & \text{if } |x_1 - w| \in E \text{ and } x_1 < w \\ \text{undefined} & \text{otherwise} \end{cases} : \mathbb{N}^2[1] \times \mathbb{N} \to \mathbb{N}^2[2].$$

Before establishing that under certain conditions minimization is computable, we need an additional concept.

Definition 1.16. An *enumeration of* a set C of type n *over* W is a map $h : \mathbb{N} \to W^n$ such that range $h = C$.

Example 1.24. Let $E \subseteq \mathbb{N}$ be the set of even natural numbers. Then the map

$$h : \mathbb{N} \to \mathbb{N}$$

such that $h(x) = 2x$ is an enumeration of E over $\mathbb{N}$.

Proposition 1.6. Let $f : C_1 \rightharpoonup C_2$ be a computable function where $C_1 \subseteq W_1^{m+1}$, $C_2 \subseteq W_2^n$, $w \in C_2$ and $\preceq$ a well-founded total order in $C_1[m+1]$. Assume that $\chi_{C_2 : W_2}$ is a computable map and $\lambda\, k . k_{(C_1[m+1], \preceq)}$ is a computable enumeration of $C_1[m+1]$. Then, $\min_{fw}^{\preceq} : C_1[1, m] \rightharpoonup C_1[m+1]$ is a computable function.

Proof. Let P_f be a program that computes f, $P_\preceq$ be an algorithm that computes $\lambda\, k . k_{(C_1[m+1], \preceq)}$ and $P_{\chi_{C_2 : W_2}}$ an algorithm that computes $\chi_{C_2 : W_2}$. We only consider the case where $m \neq 0$ since it is clear how to adapt the proof below when $m = 0$. Then, the following program P:

```
function (v₁,...,vₘ) (
    k = 0;
    b = true;
    while b do
        if Pχ_{C₂:W₂} (P_f(v₁,...,vₘ,P_≼(k))) == 1 then
            if P_f(v₁,...,vₘ,P_≼(k)) ≠ w then k = k+1 else b = false
        else
            while true do null;
        return P_≼(k)
)
```

computes $\min_{fw}^{\preceq}$ as we now show:

(1) Assume $(v_1, \ldots, v_m) \in \operatorname{dom} \min_{fw}^{\preceq}$. Let $x \in C_1[m+1]$ be such that

$$\min_{fw}^{\preceq}(v_1, \ldots, v_m) = x.$$

Hence,

$$f(v_1, \ldots, v_m, x) = w, \quad (v_1, \ldots, v_m, x') \in \operatorname{dom} f \text{ and } f(v_1, \ldots, v_m, x') \neq w \text{ for } x' \prec x.$$

Let $a \in \mathbb{N}$ be such that $a_{(C_1[m+1], \preceq)} = x$. Thus, the execution of P on $v_1, \ldots, v_m$ is such that for each $k < a$ the guards of the first and of the second if are true. When the value of k is a then the guard of the first if is true and the guard of second if is false and so P returns $a_{(C_1[m+1], \preceq)}$ after a finite number of steps.

(2) Assume $(v_1,\ldots,v_m) \notin \operatorname{dom} \min_{\overrightarrow{fw}}^{\preceq}$. Then, there are two cases.

(a) There is $x' \in C_1[m+1]$ such that $(v_1,\ldots,v_m,x') \notin \operatorname{dom} f$ and, for each $z \prec x'$, $(v_1,\ldots,v_m,z) \in \operatorname{dom} f$ and $f(v_1,\ldots,v_m,z) \neq w$. Let $a \in \mathbb{N}$ be such that $a_{(C_1[m+1],\preceq)} = x'$. Thus, the execution of P on $v_1,\ldots,v_m$ is such that for each $k < a$ the guards of the first and of the second if are true. When the value of k is a then the execution P_f in the guard of the first if either does not terminate or it does with a result not in C_2. In both cases the execution of P does not terminate.

(b) For each $z \in C_1[m+1]$, $(v_1,\ldots,v_m,z) \in \operatorname{dom} f$ and $f(v_1,\ldots,v_m,z) \neq w$. Therefore, the execution P on $v_1,\ldots,v_m$ does not terminate since the while runs forever. □

The following result is a sufficient condition for the inverse of a computable map to be a computable map using minimization.

Definition 1.17. Let $f : C_1 \rightharpoonup C_2$ be a function. The *inverse image* of f is the map

$$f^{-1} : C_2 \to \wp C_1$$

such that $f^{-1}(u) = \{w \in \operatorname{dom} f : f(w) = u\}$. When f is injective we assume that

$$f^{-1} : C_2 \rightharpoonup C_1$$

such that

$$f^{-1}(u) = \begin{cases} w & \text{whenever } u \in \operatorname{range} f \text{ and } f(w) = u \\ \text{undefined} & \text{otherwise} \end{cases}.$$

Proposition 1.7. Let $f : C_1 \to C_2$ be an injective computable map such that χ_{C_1} is a computable map and h a computable enumeration of C_1. Then,

$$f^{-1} : C_2 \rightharpoonup C_1$$

is a computable function.

Proof. (1) We start by proving that $f^{-1} = h \circ \min_{\overrightarrow{f \circ h}}^{\leq \mathbb{N}}$. Let $y \in C_2$. Consider two cases:

(a) $y \notin \operatorname{range} f$. Then $f^{-1}(y)$ is undefined. On the other hand, $\min_{\overrightarrow{f \circ h}}^{\leq \mathbb{N}}$ is also undefined since there is no $k \in \mathbb{N}$ such that $f(h(k)) = y$.

(b) $y \in \operatorname{range} f$. Then

$$f^{-1}(y) = x, \text{ where } x \text{ is the unique element of } C_1 \text{ such that } f(x) = y.$$

Moreover, since, by hypothesis, h is an enumeration (i.e. an onto map), there is $\ell \in \mathbb{N}$ such that

$$f^{-1}(y) = h(\ell).$$

Hence,

$$f^{-1}(y) = h(k), \text{ where } k \in \mathbb{N} \text{ is the smallest element such that } f(h(k)) = y.$$

Consider the following parameterized minimization:

$$\min_{f \circ h}^{\leq \mathbb{N}} = \lambda y . \min_{f \circ h y}^{\leq \mathbb{N}}(\emptyset) = \lambda y . (\mu k . f(h(k)) = y) : C_2 \to \mathbb{N}.$$

Then,

$$k = \min_{f \circ h}^{\leq \mathbb{N}}(y),$$

and so

$$f^{-1}(y) = h(k) = (h \circ \min_{f \circ h}^{\leq \mathbb{N}})(y).$$

(2) f^{-1} is a computable function.

Since f and h are computable maps then the composition $f \circ h$ is also a computable map, by Proposition 1.2. Hence, by Proposition 1.6, $\min_{f \circ h}^{\leq \mathbb{N}}$ is a computable function. Therefore, $h \circ \min_{f \circ h}^{\leq \mathbb{N}}$ is computable by Proposition 1.2 and so f^{-1} is computable by Proposition 1.1. □

Remark 1.9. Taking into account that A is a finite set, we fix an order denoted by

$$\prec_\mathsf{A}$$

satisfying the following properties:

- *totality*, that is, for every $a, b \in \mathsf{A}$, we either have $a \prec_\mathsf{A} b$ or $a = b$ or $b \prec_\mathsf{A} a$;

- *irreflexivity*, that is, for every $a \in \mathsf{A}$, it is not the case that $a \prec_\mathsf{A} a$;

- *asymmetry*, that is, for every $a, b \in \mathsf{A}$, if $a \prec_\mathsf{A} b$ then it is not the case $b \prec_\mathsf{A} a$;

- *transitivity*, that is, for every $a, b, c \in \mathsf{A}$, if $a \prec_\mathsf{A} b$ and $b \prec_\mathsf{A} c$ then $a \prec_\mathsf{A} c$.

Thus $\prec_\mathsf{A}$ is a strict total order. Using $\prec_\mathsf{A}$, the objective now is to prove that there exists an injective enumeration for W.

Proposition 1.8. The order $\prec_\mathsf{A}$ induces an injective enumeration

$$\lambda k . k_\mathsf{W} : \mathbb{N} \to \mathsf{W}$$

of W.

Proof. Consider the function $S_\mathsf{A} : \mathsf{A} \rightharpoonup \mathsf{A}$ such that $S_\mathsf{A}(a)$ is the least of all elements greater than a, whenever these exist. For each $m \in \mathbb{N}$, S_A is extended to the set of words of length m using the lexicographic order. Finally, it is extended to W by requiring that, for each m, the last word of length m be followed by the first word of length $m+1$. The envisaged enumeration $\lambda k . k_\mathsf{W}$ is the map that for each k returns the k-th element of W as follows: 0_W is the empty sequence and $(k+1)_\mathsf{W}$ is $S_\mathsf{W}(k_\mathsf{W})$. By construction, this map is injective and an enumeration of universe W. □

Observe that the proposed injective enumeration follows the lexicographic order of W induced by the given ordering of A.

Remark 1.10. We denote by

$$J_W : W \to \mathbb{N}$$

the inverse of the bijection $\lambda k . k_W$.

We now introduce a total and well-founded relation $\preceq_W$ over W.

Definition 1.18. The order $\preceq_W$ over W is as follows:

$$w_1 \preceq_W w_2 \quad \text{if} \quad J_W(w_1) \leq J_W(w_2).$$

Observe that the first element 0_W in this ordering is ε.

Exercise 1.7. Show that the order $\preceq_W$ is well-founded and total.

Observe that given $W \subseteq$ W, the restriction of $\preceq_W$ to W is also total and well-founded.

Exercise 1.8. Let W be a working universe. Define bijections

$$\lambda k . k_W : \mathbb{N} \to W$$

and

$$J_W : W \to \mathbb{N}$$

such that they are inverse of each other. Moreover, show that there are computable bijections when W is the working universe W.

Recursive functions

Capitalizing on the results of this section we now establish that each recursive function in the sense of Stephen Kleene ([33, 26, 35]) is computable in our model.

Definition 1.19. The set $\mathscr{R}$ of recursive functions over the working universe $\mathbb{N}$ is defined inductively as follows:

- $\lambda . k : \{\emptyset\} \to \mathbb{N}$ is in $\mathscr{R}$ for each $k \in \mathbb{N}$;

- $\lambda k . 0 : \mathbb{N} \to \mathbb{N}$ is in $\mathscr{R}$;

- $\lambda k . k + 1 : \mathbb{N} \to \mathbb{N}$ is in $\mathscr{R}$;

- $\mathrm{proj}_{[i]}^{\mathbb{N}^n} : \mathbb{N}^n \to \mathbb{N}$ is in $\mathscr{R}$ for every $n \in \mathbb{N}^+$ and $1 \leq i \leq n$;

- if $f : \mathbb{N}^k \rightharpoonup \mathbb{N}^m$ and $g : \mathbb{N}^m \rightharpoonup \mathbb{N}^n$ are in $\mathscr{R}$ then $g \circ f : \mathbb{N}^k \rightharpoonup \mathbb{N}^n$ is in $\mathscr{R}$;

- if $f : \mathbb{N}^{n+1} \rightharpoonup \mathbb{N}$ is in $\mathscr{R}$ then $\min_{f0}^{\leq \mathbb{N}} : \mathbb{N}^n \rightharpoonup \mathbb{N}$ is in $\mathscr{R}$;

- if $f : \mathbb{N}^n \rightharpoonup \mathbb{N}$ and $g : \mathbb{N}^{n+2} \rightharpoonup \mathbb{N}$ are in $\mathscr{R}$ then $\mathrm{rec}(f,g) : \mathbb{N}^{n+1} \rightharpoonup \mathbb{N}$ defined as follows

  - $\mathrm{rec}(f,g)(k_1,\ldots,k_n,0) = f(k_1,\ldots,k_n)$
  - $\mathrm{rec}(f,g)(k_1,\ldots,k_n,k+1) = g(k_1,\ldots,k_n,k,\mathrm{rec}(f,g)(k_1,\ldots,k_n,k))$

is in $\mathscr{R}$.

Proposition 1.9. Every recursive function is computable.

Proof. The proof is by induction on the construction of a recursive function in $\mathscr{R}$.

(Base) The first three cases are straightforward. The projections are computable according to Proposition 1.5.

(Step) Observe that $\chi_{\mathbb{N}^n}$ is a computable function for each $n \in \mathbb{N}^+$ (see Exercise 1.3).

(Composition) Let $f : \mathbb{N}^k \rightharpoonup \mathbb{N}^m$ and $g : \mathbb{N}^m \rightharpoonup \mathbb{N}^n$ be recursive functions. Then, by the induction hypothesis they are computable and so by Proposition 1.2, $g \circ f$ is a computable function.

(Minimization) Let $f : \mathbb{N}^{n+1} \rightharpoonup \mathbb{N}$ be a recursive function. Then f is computable and so $\min_{f0}^{\leq \mathbb{N}}$ is also computable by Proposition 1.6.

(Recursion) Let $f : \mathbb{N}^n \rightharpoonup \mathbb{N}$ and $g : \mathbb{N}^{n+2} \rightharpoonup \mathbb{N}$ be recursive functions. Then f and g are computable functions. Let P_f and P_g be programs that compute f and g, respectively. Then the program

```
function (k₁,...,kₙ,k) (
    if k == 0 then
        return Pf(k₁,...,kₙ);
    r = Pf(k₁,...,kₙ);
    j = 0;
    while isnat(r) ∧ j < k do (
        r = Pg(k₁,...,kₙ,j,r);
        j = j+1
    )
    return r
)
```

computes $\mathrm{rec}(f,g)$. Observe that when return $P_f(k_1,\ldots,k_n)$ is executed then the program terminates. □

1.5   Solved Problems and Exercises

Problem 1.1. Let $f : C_1 \rightharpoonup C_2$. Show that,

- if $\mathrm{dom}\, f$ is finite, then f is computable;

- if C_1 is finite, then f is computable.

Solution Assume that $\mathrm{dom}\, f$ is finite and let $\mathrm{dom}\, f = \{c_1, \ldots, c_n\}$. We know that $f(\mathrm{dom}\, f) = \{f(c_1), \ldots, f(c_n)\}$. Hence, the following procedure computes f

```
function (w) (
    if w == c₁ then return f(c₁)
    else if w == c₂ then return f(c₂)
        ⋮
    else if w == cₙ then return f(cₙ)
    else while true do null
)
```

If C_1 is finite then, as $\mathrm{dom}\, f \subseteq C_1$, the result follows from the previous result. ∇

Problem 1.2. Show that the following function:

$$\mathrm{even} = \lambda\, k\, . \begin{cases} 1 & \text{if } k \text{ is even} \\ \text{undefined} & \text{otherwise} \end{cases} : \mathbb{N} \rightharpoonup \mathbb{N}.$$

is computable defining even using an appropriate minimization.

Solution Consider the function

$$g = \lambda\, k_1, k_2 . |k_1 - 2k_2| : \mathbb{N}^2 \to \mathbb{N}.$$

Observe that

$$g = (\lambda\, x_1, x_2 . |x_1 - x_2|) \circ \langle \mathrm{proj}_{[1]}^{\mathbb{N}^2}, (\lambda\, x, y . x \times y) \circ \langle (\lambda\, x . 2) \circ \mathrm{proj}_{[1]}^{\mathbb{N}^2}, \mathrm{proj}_{[2]}^{\mathbb{N}^2} \rangle \rangle.$$

It is immediate to find programs that compute $\lambda\, x . 2$, $\lambda\, x, y . x \times y$ and $\lambda\, x_1, x_2 . |x_1 - x_2|$. Moreover, $\mathrm{proj}_{[i]}^{\mathbb{N}^2}$ for $i = 1, 2$ are computable by Proposition 1.5. Then, by Proposition 1.2 and by Proposition 1.4, we conclude that g is computable. On the other hand,

$$\mathrm{even} = (\lambda\, x . 1) \circ \min_{g0}^{\leq \mathbb{N}}.$$

Therefore, even is computable by Proposition 1.6. ∇

Problem 1.3. For each $n \in \mathbb{N}^+$ show that the following maps are bijections between $\mathbb{N}^n$ and $\mathbb{N}$ and, furthermore, inverse of each other:

- $\beta_1^n : \mathbb{N} \to \mathbb{N}^n$ inductively defined as follows:

  - $\beta_1^1(x) = x$ for each $x \in \mathbb{N}$;
  - $\beta_1^n(x) = (K(x), \beta_1^{n-1}(L(x)))$ $x \in \mathbb{N}$.

- $\beta_n^1 : \mathbb{N}^n \to \mathbb{N}$ inductively defined as follows:

  - $\beta_1^1(x) = x$ for each $x \in \mathbb{N}$;
  - $\beta_n^1(x_1,\ldots,x_n) = J(x_1, \beta_{n-1}^1(x_2,\ldots,x_n))$ for $x_1,\ldots,x_n \in \mathbb{N}$.

Solution We prove by induction on n that β_1^n is a bijection.

(Base) $n = 1$. The map β_1^1 is injective and surjective since it is the identity over $\mathbb{N}$.

(Step) $n > 1$.

(1) Surjective. Let $(x_1,\ldots,x_n) \in \mathbb{N}^n$. Observe that

$$(x_1,\ldots,x_n) = (x_1,(x_2,\ldots,x_n)).$$

By the induction hypothesis, there is x' such that $\beta_1^{n-1}(x') = (x_2,\ldots,x_n)$. Take

$$x = J(x_1,x').$$

Then
$$
\begin{aligned}
\beta_1^n(x) &= (K(J(x_1,x')), \beta_1^{n-1}(L(J(x_1,x')))) \\
&= (x_1, \beta_1^{n-1}(x')) \\
&= (x_1,(x_2,\ldots,x_n)) \\
&= (x_1,\ldots,x_n)
\end{aligned}
$$

(2) Injective. Let $x \neq y$ with $x,y \in \mathbb{N}$.

(a) $K(x) \neq K(y)$. Then, $\beta_1^n(x) \neq \beta_1^n(y)$.

(b) $K(x) = K(y)$. Then, there are pairs (x_1,x_2) and (y_1,y_2) such that

$$J(x_1,x_2) = x \text{ and } J(y_1,y_2) = y$$

where $(x_1,x_2) \neq (y_1,y_2)$ since J is a bijection, see Exercise 1.1. Observe that $x_1 = K(x)$, $x_2 = L(x)$, $y_1 = K(y)$ and $y_2 = L(y)$. So $x_2 = L(x) \neq L(y) = y_2$. Hence, by the induction hypothesis,

$$\beta_1^{n-1}(L(x)) \neq \beta_1^{n-1}(L(y)).$$

So, $\beta_1^n(x) \neq \beta_1^n(y)$.

We prove by induction on n that β_n^1 is a bijection.

(Base) $n = 1$. Th map β_1^1 is injective and surjective since it is the identity over $\mathbb{N}$.

(Step) $n > 1$.

(1) Surjective. Let $x \in \mathbb{N}$. Then, there are x_1, x_2 such that $J(x_1, x_2) = x$. Let $y_2, \ldots, y_n$ be such that

$$x_2 = \beta^1_{n-1}(y_2, \ldots, y_n)$$

which exists since, by the induction hypothesis, β^1_{n-1} is surjective. Hence

$$(x_1, (y_2, \ldots, y_n))$$

is such that

$$\beta^1_n(x_1, (y_2, \ldots, y_n)) = J(x_1, \beta^1_{n-1}(y_2, \ldots, y_n)) = x.$$

(2) Injective. Straightforward.

Finally, we prove by induction that β^n_1 and β^1_n are inverse of each other. The base case is straightforward. The step is as follows:

$$\begin{aligned}
\beta^n_1(\beta^1_n(x_1, \ldots, x_n)) &= \beta^n_1(J(x_1, \beta^1_{n-1}(x_2, \ldots, x_n))) \\
&= (K(J(x_1, \beta^1_{n-1}(x_2, \ldots, x_n)), \\
&\qquad \beta^{n-1}_1(L(J(x_1, \beta^1_{n-1}(x_2, \ldots, x_n)))))) \\
&= (x_1, \beta^{n-1}_1(\beta^1_{n-1}(x_2, \ldots, x_n))) \\
&= (x_1, \ldots, x_n).
\end{aligned}$$

$\triangledown$

Problem 1.4. Show that β^n_1 and β^1_n are computable maps for every $n \in \mathbb{N}^+$.

Solution Let P_K, P_L and P_J be algorithms that compute K, L and J, respectively (see Exercise 1.2). We show, by induction on $n \in \mathbb{N}^+$ that β^n_1 is computable.

(Base) $n = 1$. It is immediate to see that

$$\text{function } (k) \ (\text{return } k)$$

computes β^1_1.

(Step) $n > 1$. Assume, by induction hypothesis, that β^{n-1}_1 is a computable map. Let $P_{\beta^{n-1}_1}$ be an algorithm that computes β^{n-1}_1. Then the following algorithm

$$\text{function } (k) \ (\text{return } P_K(k), P_{\beta^{n-1}_1}(P_L(k)))$$

computes β^n_1. The proof that β^1_n is computable also follows by a straightforward induction. $\triangledown$

Problem 1.5. For $m, n \in \mathbb{N}^+$, show that $f : \mathbb{N}^n \rightharpoonup \mathbb{N}^m$ is computable if and only if

$$\beta^1_m \circ f \circ \beta^n_1 : \mathbb{N} \rightharpoonup \mathbb{N}$$

is computable.

Solution ($\rightarrow$) Assume that f is computable. Since β_m^1 and β_1^n are computable (see Problem 1.4) then $\beta_m^1 \circ f \circ \beta_1^n$ is also computable by Proposition 1.2 because $\chi_{\mathbb{N}}$ is a computable map.

($\leftarrow$) Assume that $\beta_m^1 \circ f \circ \beta_1^n : \mathbb{N} \rightarrow \mathbb{N}$ is a computable function. Observe that

$$\beta_1^m \circ (\beta_m^1 \circ f \circ \beta_1^n) \circ \beta_n^1 = (\beta_1^m \circ \beta_m^1) \circ f \circ (\beta_1^n \circ \beta_n^1) = f$$

by Problem 1.3. Therefore, f is a computable function by Proposition 1.2 since it is the composition of computable functions and $\chi_{\mathbb{N}}$ is a computable map. ∇

Supplementary Exercises

Exercise 1.9. Show that if $w \notin \text{range} f$ then $\min_{fw}^{\preceq}$ is $\lambda x.\text{undefined}$. Does the converse hold?

Exercise 1.10. Let

$$f = \lambda x. \begin{cases} \dfrac{x}{3} & \text{if } x \text{ is a multiple of 3} \\ \text{undefined} & \text{otherwise} \end{cases} : \mathbb{N} \rightarrow \mathbb{N}.$$

Show how to obtain f from function $\lambda x_1, x_2 . |x_1 - x_2| : \mathbb{N}^2 \rightarrow \mathbb{N}$. Conclude that f is computable.

Exercise 1.11. Let $f = \lambda x_1, x_2, x_3 . (x_1^2 + x_3) \times x_2 : \mathbb{N}^3 \rightarrow \mathbb{N}$. Show how to obtain f from functions $\lambda x_1, x_2 . x_1 + x_2 : \mathbb{N}^2 \rightarrow \mathbb{N}$ and $\lambda x_1, x_2 . x_1 \times x_2 : \mathbb{N}^2 \rightarrow \mathbb{N}$. Conclude that f is computable.

Exercise 1.12. Let $f = \lambda x . x^2 : \mathbb{N} \rightarrow \mathbb{N}$. Show how to obtain f from function $\lambda x_1, x_2 . x_1^{x_2} : \mathbb{N}^2 \rightarrow \mathbb{N}$. Conclude that f is computable.

Chapter 2

Decidable and Listable Sets

In this chapter we introduce the important notions of decidable set and listable set. Moreover, we analyze the closure of the classes of decidable sets and listable sets under set operations as well as preservation and reflection of decidability and listability by computable functions.

2.1 Decidable Sets

Intuitively, a set C of type n over W is decidable if, given $w \in W^n$, it is possible to decide, in a finite amount of time, whether w belongs to C or to its complement.

Definition 2.1. Let W be a working universe. A set C is said to be *decidable on W* if it is a set over W and its characteristic map $\chi_{C:W}$ is computable.

That is, there is an algorithm capable of computing the characteristic map $\chi_{C:W}$ that when executed on $w \in W^n$ terminates either returning 1 (positive answer), in which case $w \in C$, or returning 0 (negative answer), in which case $w \in W^n \setminus C$.

We say that C is *decidable* when W is W. In this case, we may write C^{c} for $W^n \setminus C$.

Example 2.1. Let M_8 be the set of natural numbers that are multiples of 8. It is immediate to see that the algorithm

$$\text{function } (w) \ (\text{if } w \% 8 == 0 \text{ then return 1 else return 0})$$

computes $\chi_{M_8:\mathbb{N}}$. Then, by definition, M_8 is decidable.

Exercise 2.1. Show that W and $\mathbb{N}$ are decidable sets.

It seems interesting to ask some questions about classes of decidable sets, like if the finite union of decidable sets is still a decidable set. The following proposition provides a first answer.

31

Proposition 2.1. Each class of decidable sets on W of the same type is closed under complement and intersection.

Proof. (1) Closure under complement. Let $C \subseteq W^n$ be a decidable set. The algorithm

$$\text{function } (w) \; (\text{return } 1 - P_{\chi_{C:W}}(w))$$

computes $\chi_{(W^n \setminus C):W}$ where $P_{\chi_{C:W}}$ is an algorithm that computes $\chi_{C:W}$. Hence, $W^n \setminus C$ is a decidable set.

(2) Closure under intersection. Let $C_1, C_2 \subseteq W^n$ be decidable sets. The algorithm

$$\text{function } (w) \; (\text{return } P_{\chi_{C_1:W}}(w) \times P_{\chi_{C_2:W}}(w))$$

computes $\chi_{C_1 \cap C_2:W}$ where $P_{\chi_{C_i:W}}$ is an algorithm that computes $\chi_{C_i:W}$ for $i = 1, 2$. Hence, $C_1 \cap C_2$ is a decidable set. $\square$

Exercise 2.2. State and prove a similar result concerning closure under union and Cartesian product.

Exercise 2.3. Show that there are precisely $|\mathbb{N}|$ decidable sets over $\mathbb{N}$.

(So, since $|\wp\mathbb{N}| > |\mathbb{N}|$, there are subsets of $\mathbb{N}$ that are undecidable.)

In what concerns functions it seems worthwhile to ask about preservation and reflection results like is decidability preserved/reflected by computable functions? The following result is concerned with the decidability of back images of decidable sets given by computable functions (reflection result).

Proposition 2.2. Let C_1 and C_2 be sets over W_1 and W_2, respectively, $f : C_1 \rightharpoonup C_2$ a computable function such that the domain is a decidable set on W_1, and $B \subseteq C_2$. If B is decidable on W_2, then

$$f^{-1}(B)$$

is decidable on W_1.

Proof. Let P_f be a program that computes f. Moreover, let $P_{\chi_{\text{dom} f:W_1}}$ and $P_{\chi_{B:W_2}}$ be algorithms for $\chi_{\text{dom} f:W_1}$ and $\chi_{B:W_2}$, respectively, which exist since, by hypothesis, f has a decidable domain on W_1 and B is a decidable set on W_2. Consider the following algorithm P:

$$\text{function } (w) \; (\text{if } P_{\chi_{\text{dom} f:W_1}}(w) == 0 \text{ then return } 0 \text{ else return } P_{\chi_{B:W_2}}(P_f(w)))$$

We now show that P computes $\chi_{f^{-1}(B):W_1}$. Assume that C_1 is of type n. Let $a \in (W_1)^n$.

(1) Assume that $a \in f^{-1}(B)$. Then, $a \in \text{dom} f$ and $f(a) \in B$. That is, $\chi_{\text{dom} f:W_1}(a) = 1$ and $\chi_{B:W_2}(f(a)) = 1$. Therefore, in the execution of P on a the guard of the if is false and so P executed on a terminates with 1 which is $\chi_{B:W_2}(f(a))$.

(2) Assume that $a \notin f^{-1}(B)$. That is, either $a \notin \operatorname{dom} f$ or $f(a) \notin B$. Consider two cases: (a) $a \notin \operatorname{dom} f$. Thus, the execution of $P_{\chi_{\operatorname{dom} f:W_1}}$ on a returns 0 and so the execution of P on a returns 0. (b) $f(a) \notin B$. In this case, the execution of $P_{\chi_{B:W_2}}$ on $f(a)$ returns 0 and so the execution of P on a returns 0. $\qquad\square$

On the other hand, the image of a decidable set given by a computable function (or even by a computable map) with decidable domain is not necessarily decidable. In fact, as will be seen later (in Section 3.2), there exists a non-decidable set enumerated by a computable map. Let C be such a set (necessarily infinite), enumerated by a computable map

$$h : \mathbb{N} \to C.$$

Clearly, map h provides the envisaged counterexample: $\mathbb{N}$ is decidable, but $h(\mathbb{N}) = C$ is not. Sets that have computable enumerations are very important in the development of Computability Theory. They are analyzed in the next section.

2.2 Listable Sets

Intuitively speaking, a set C is listable if, given $w \in C$, it is possible to ascertain this fact in a finite amount of time.

Definition 2.2. Let W be a working universe. A set is said to be *listable* (or *recursively enumerable* or *semidecidable*) on W if it is a set over W and either it is empty or it allows a computable enumeration over W. When W is W, we say that the set is *listable*.

Indeed, to this end, an algorithm for computing an enumeration h of $C \neq \emptyset$ can be used as follows. Check if $h(i) = w$ for consecutive values of i starting from 0. If $w \in C$ then, after a finite number of attempts, a k will be found such that $h(k) = w$. Clearly, this search will not terminate if $w \notin C$.

It seems interesting to ask some questions about classes of listable sets, like if the finite union of listable sets is still a listable set. The following proposition provides a first answer.

Proposition 2.3. Each class of listable sets on W of the same type is closed under union.

Proof. Let C_1 and C_2 be listable sets on W of type n. If $C_1 = \emptyset$ then $C_1 \cup C_2 = C_2$ which is listable on W. Similarly, for $C_2 = \emptyset$. Finally, assume that C_1 and C_2 are both non-empty. Let h_1 and h_2 be enumerations of C_1 and C_2 over W computed by algorithms P_{h_1} and P_{h_2}, respectively. Let

$$h = \lambda x. \begin{cases} h_1\left(\dfrac{x}{2}\right) & \text{if } x \text{ is even} \\ h_2\left(\dfrac{x-1}{2}\right) & \text{otherwise} \end{cases} : \mathbb{N} \to W^n.$$

We start by showing that h is an enumeration of $C_1 \cup C_2$ over W. Let $c \in C_1 \cup C_2$. If $c \in C_1$ then there exists $x_1 \in \mathbb{N}$ such that $h_1(x_1) = c$. Choosing $x = 2x_1$ it follows that

$$h(x) = h_1\left(\frac{2x_1}{2}\right) = c.$$

If $c \in C_2$ then there exists $x_2 \in \mathbb{N}$ such that $h_2(x_2) = c$. Choosing $x = 2x_2 + 1$ it follows that

$$h(x) = h_2\left(\frac{(2x_2+1)-1}{2}\right) = c.$$

Therefore, h is an enumeration for $C_1 \cup C_2$. Moreover, the following algorithm

$$\text{function } (k) \; (\text{if } k\%2 == 0 \text{ then } P_{h_1}(k/2) \text{ else } P_{h_2}((k-1)/2))$$

computes h. $\qquad\square$

Proposition 2.4. The class of listable sets on W is closed under Cartesian product.

Proof. Let C be a listable set on W of type m. If $C = \emptyset$ then $C^n = \emptyset$ which is listable on W, for $n \in \mathbb{N}^+$. Otherwise, assume that C is non-empty. Let $h : \mathbb{N} \to W^m$ be an enumeration of C over W computed by the algorithm P_h. We now show that there is a computable enumeration of C^n over W. If $n = 1$ then h is a a computable enumeration of C^1 over W. Otherwise consider the function:

$$h_{C^n} = \langle h \circ \text{proj}_{[1]}^{\mathbb{N}^n} \circ \beta_1^n, \ldots, h \circ \text{proj}_{[n]}^{\mathbb{N}^n} \circ \beta_1^n \rangle : \mathbb{N} \to W^{mn}.$$

Then:

(a) h_{C^n} is a map since it is the aggregation of n compositions of maps;

(b) range $h_{C^n} = C^n$. Let $(c_1, \ldots, c_n) \in C^n$. Then, there are $x_1, \ldots, x_n \in \mathbb{N}$ such that

$$h(x_i) = c_i$$

for each $i = 1, \ldots, n$. Let $x \in \mathbb{N}$ be such that $x = \beta_n^1(x_1, \ldots, x_n)$. Hence,

$$h_{C^n}(x) = \langle h \circ \text{proj}_{[1]}^{\mathbb{N}^n} \circ \beta_1^n, \ldots, h \circ \text{proj}_{[n]}^{\mathbb{N}^n} \circ \beta_1^n \rangle(x)$$

$$= ((h \circ \text{proj}_{[1]}^{\mathbb{N}^n} \circ \beta_1^n)(x), \ldots, (h \circ \text{proj}_{[n]}^{\mathbb{N}^n} \circ \beta_1^n)(x))$$

$$= ((h \circ \text{proj}_{[1]}^{\mathbb{N}^n})(\beta_1^n(x)), \ldots, (h \circ \text{proj}_{[n]}^{\mathbb{N}^n})(\beta_1^n(x)))$$

$$= ((h \circ \text{proj}_{[1]}^{\mathbb{N}^n})(x_1, \ldots, x_n), \ldots, (h \circ \text{proj}_{[n]}^{\mathbb{N}^n})(x_1, \ldots, x_n))$$

$$= (h(\text{proj}_{[1]}^{\mathbb{N}^n}(x_1, \ldots, x_n)), \ldots, h(\text{proj}_{[n]}^{\mathbb{N}^n}(x_1, \ldots, x_n)))$$

$$= (h(x_1), \ldots, h(x_n))$$

$$= (c_1, \ldots, c_n)$$

and so the thesis follows.

(c) h_{C^n} is computable. Let $P_{\beta_1^n}$ be an algorithm that computes β_1^n (see Problem 1.4). The following program:

```
function (k) (
    x_1,...,x_n = P_{β_1^n}(k);
    return P_h(x_1),...,P_h(x_n)
)
```

computes h_{C^n}.   □

Exercise 2.4. Show that each class of listable sets on W of the same type is closed under intersection. Additionally, generalize Proposition 2.4 to arbitrary sets.

We now show that the class of decidable sets is contained in the class of listable sets.

Proposition 2.5. Let W be a listable set. Then, every decidable set on W is also listable on W.

Proof. Let C be a decidable set of type n over W. If C is empty, then C is listable on W by definition. Otherwise, let $c \in C$. Furthermore, since in this case $W \neq \emptyset$, let h be a computable enumeration of W^n (observe that if W is listable so is W^n by Proposition 2.4). Let $P_{\chi_{C:W}}$ and P_h be algorithms that compute $\chi_{C:W}$ and h, respectively. Then, the following algorithm:

```
function (k) (if P_{χ_{C:W}} (P_h(k)) == 1 then return P_h(k) else return c)
```

computes an enumeration of C. Hence, C is a listable set on W.   □

The converse of this result does not hold in general, as it will be seen in Section 3.2, where examples are provided of listable sets on $\mathbb{N}$ that are non-decidable sets on $\mathbb{N}$.

Example 2.2. Recall the decidable set M_8 on $\mathbb{N}$ defined in Example 2.1. Then, by Proposition 2.5, M_8 is also a listable set on $\mathbb{N}$.

The following result provides a characterization of decidable sets in terms of listable sets.

Proposition 2.6 (Post's Theorem). Let W be a working universe and C a set of type n over W. If C and $W^n \setminus C$ are both listable sets on W, then both are decidable sets on W.

Proof. If either C or $W^n \setminus C$ is empty, then the result is trivial. Otherwise, let $h_C : \mathbb{N} \to W^n$ and $h_{W^n \setminus C} : \mathbb{N} \to W^n$ be computable enumerations of C and $W^n \setminus C$ over W, respectively. Let P_{h_C} and $\mathsf{P}_{h_{W^n \setminus C}}$ be algorithms that compute h_C and $h_{W^n \setminus C}$, respectively. Then, the following algorithm computes the characteristic map $\chi_{C:W}$:

```
function (w) (
    k = 0;
    while P_hc(k) ≠ w ∧ P_hwⁿ\c(k) ≠ w do
        k = k + 1;
    if P_hc(k) == w then
        return 1
    else
        return 0
)
```

Therefore, C is a decidable set on W and by Proposition 2.1 $W^n \setminus C$ is also a decidable set on W. $\qquad\square$

In particular, if C and C^c are both listable then they are both decidable (the traditional statement of Post's Theorem). Observe also that Post's Theorem entails the non-listability of the complement of a listable but non-decidable set.

The following result is concerned with the listability of images of listable sets given by computable functions.

Proposition 2.7. Let C_1 and C_2 be sets over W_1 and W_2, respectively, $f : C_1 \rightharpoonup C_2$ a computable function and $A \subseteq C_1$. If A is listable on W_1 and C_2 is a decidable set, then $f(A \cap \operatorname{dom} f)$ is listable on W_2.

Proof. Clearly, $f(A \cap \operatorname{dom} f)$ is a set over W_2. If $f(A \cap \operatorname{dom} f) = \emptyset$ then $f(A \cap \operatorname{dom} f)$ is listable on W_2 by definition. If $f(A \cap \operatorname{dom} f) \neq \emptyset$ then $(A \cap \operatorname{dom} f) \neq \emptyset$ and, so, $A \neq \emptyset$. In this case, let $h : \mathbb{N} \to W_1^n$ be a computable enumeration of A over W_1 (assuming that C_1 is of type n) and $a \in (A \cap \operatorname{dom} f)$. Let P_f, P_h, $\mathsf{P}_{\chi_{C_2}}$, P_K and P_L be programs that compute f, h, χ_{C_2}, K and L, respectively. Thus, the following algorithm computes an enumeration of $f(A \cap \operatorname{dom} f)$ over W_2:

```
function (k) (
    i = P_K(k);
    t = P_L(k);
    y = stceval(P_f, P_h, i, t, P_f(a));
    if P_χC₂(y) == 1 then return y else return P_f(a)
)
```

Therefore, if the execution of P_h on i plus the execution of P_f on $h(i)$ terminates with a value in C_2 within t units of time, then the output is $f(h(i))$. Otherwise, the result is $f(a)$. Thus, in both cases, the result is in $f(A \cap \operatorname{dom} f)$ and, so, the range of the map computed by this algorithm is contained in $f(A \cap \operatorname{dom} f)$. Furthermore, that range coincides with $f(A \cap \operatorname{dom} f)$, as required, because for any $b \in f(A \cap \operatorname{dom} f)$ there are i and t such that the execution of P on $J(i,t)$ is successful within t units of time with output b. Hence, $f(A \cap \operatorname{dom} f)$ is listable over W_2. □

In particular, if A is listable then $f(A \cap \operatorname{dom} f)$ is also listable. Moreover, if f is a map then $f(A)$ is listable on W_2 whenever A is listable on W_1.

What can be said about the converse of Proposition 2.7? That is, given f computable, is the assertion

$$\text{if } f(A) \text{ is a listable set then } A \text{ is a listable set}$$

true? In general it is not the case. Consider the following counterexample. Take A to be a non-listable set (wait until Chapter 3) and let $f = \lambda x . b$. Clearly, f is computable and $f(A) = \{b\}$ is listable. Indeed, $\{b\}$ is a finite set therefore is a decidable set and so, by Proposition 2.5, it is a listable set. Nevertheless, we have the following result.

Proposition 2.8. Let C_1 and C_2 be sets over W_1 and W_2, respectively, $f : C_1 \rightharpoonup C_2$ a computable function with listable domain on W_1 and $B \subseteq C_2$. If B is listable on W_2, then $f^{-1}(B)$ is listable on W_1.

Proof. If $f^{-1}(B) = \emptyset$ then $f^{-1}(B)$ is listable on W_1. If $f^{-1}(B) \neq \emptyset$ then $B \neq \emptyset$ and $\operatorname{dom} f \neq \emptyset$. In this case, let a be an element of $f^{-1}(B)$, h_B a computable enumeration of B and $h_{\operatorname{dom} f}$ a computable enumeration of $\operatorname{dom} f$. Let P_{h_B}, $P_{h_{\operatorname{dom} f}}$, P_f, P_K and P_L be programs that compute h_B, $h_{\operatorname{dom} f}$, f, K and L, respectively. Then, the following algorithm

$$
\begin{aligned}
&\text{function } (k) \ (\\
&\quad i_1 = P_K(k); \\
&\quad i_2 = P_L(k); \\
&\quad \text{if } P_f(P_{h_{\operatorname{dom} f}}(i_1)) \neq P_{h_B}(i_2) \text{ then} \\
&\qquad \text{return } a \\
&\quad \text{else} \\
&\qquad \text{return } P_{h_{\operatorname{dom} f}}(i_1) \\
&)
\end{aligned}
$$

computes an enumeration of $f^{-1}(B)$. □

The following result provides an interesting property of infinite listable sets that will be useful in the sequel.

Proposition 2.9. Every infinite listable set on W has an injective computable enumeration over W.

Proof. Let C be an infinite listable set of type n over W. Let $h : \mathbb{N} \to W^n$ be an enumeration of C over W. Let $g : \wp\mathbb{N} \setminus \{\emptyset\} \to \wp C \setminus \{\emptyset\}$ be defined as follows:

- $g(\{0\}) = \{h(0)\};$

- $g(\{0,\ldots,k+1\}) = g(\{0,\ldots,k\}) \cup$

  $\{h(m) : m$ is the smallest natural number such that $h(m) \neq h(j), j = 0,\ldots,k\}.$

Observe that $g(\{0,\ldots,k\})$ has $k+1$ elements, for every $k \in \mathbb{N}$ and, moreover,

$$g(\{0,\ldots,k_1\}) \subset g(\{0,\ldots,k_2\}), \text{ for every } k_1,k_2 \in \mathbb{N} \text{ such that } k_1 < k_2.$$

Define map

$$\bar{h} : \mathbb{N} \to W^n$$

inductively as follows:

- $\bar{h}(0)$ is the unique element in $g(\{0\});$

- $\bar{h}(k+1)$ is the unique element in $g(\{0,\ldots,k+1\}) \setminus g(\{0,\ldots,k\}).$

We now show that $\bar{h}$ is an injective enumeration of C.

(1) $\bar{h}$ is an enumeration of C. Let $c \in C$. Then, there is a smallest natural number m such that $h(m) = c$. We have two cases: (i) $m = 0$. Then, $\bar{h}(0) = h(0) = c$. (ii) $m \neq 0$. Consider the multiset $\{h(0),\ldots,h(m-1)\}$. Let $k \in \mathbb{N}$ be such that $g(\{0,\ldots,k\})$ is the set of all distinct elements of this multiset. Thus, $g(\{0,\ldots,k+1\}) = g(\{0,\ldots,k\}) \cup \{h(m)\}$. Hence, $\bar{h}(k+1) = c$.

(2) $\bar{h}$ is an injective map. Let $k_1 \neq k_2 \in \mathbb{N}$. Assume, without loss of generality, that $k_1 < k_2$. Then, $g(\{0,\ldots,k_1\}) \subset g(\{0,\ldots,k_2\})$ and so, it is easy to show that $\bar{h}(k_1) \neq \bar{h}(k_2)$ by induction on k_2.

(3) $\bar{h}$ is computable. Let P_h be an algorithm that computes h. Consider the program P

```
function (k) (
    c = 0;
    i = 0;
    while c ≠ k + 1 do (
        j = 0;
        b = false;
        while j < i ∧ b == false do
            if Pₕ(j) == Pₕ(i) then
                b = true
            else
                j = j + 1;
        if b == false then
            c = c + 1;
        if c ≠ k + 1 then
            i = i + 1
    );
    return Pₕ(i)
)
```

Observe that in P the variable c, for each natural number i, counts the number of distinct elements that occurred in the enumeration h of C up to i. Thus, the execution of P on k terminates when c is $k + 1$ returning $h(i)$. ☐

The following result establishes a sufficient condition for the listability of a countably infinite union of listable sets.

Proposition 2.10. Let $s : \mathbb{N} \to \mathbb{N}$ be a computable map and $C_i \subseteq W$ for each $i \in \mathbb{N}$. Assume that, for each $i \in \mathbb{N}$, $(s(i))_W$ is a program that computes an enumeration of C_i. Then,

$$\bigcup_{i \in \mathbb{N}} C_i$$

is a listable set.

Proof. Consider the following algorithm P:

```
function (k) (
    i = PK(k);
    j = PL(k);
    return PW(Pₛ(i))(j)
)
```

where P_K, P_L, P_s and P_W are algorithms that compute K, L, s and $\lambda k. k_W$, respectively.

Let $w \in \bigcup_{i \in \mathbb{N}} C_i$. Then, there is i such that $w \in C_i$. So, there is $j \in \mathbb{N}$ such that the execution of the algorithm $s(i)_W$ on j returns w. Let $k = J(i, j)$. Then the execution of P on k returns w.   □

The following result provides several necessary and sufficient conditions for a set to be listable.

Proposition 2.11 (Listability Criteria). Let W be a decidable universe, $m, n_1, n_2 \in \mathbb{N}$ with $n_1 > 0$, V_1 an infinite listable set on V_1, V_2 a non-empty decidable set and C a set of type m over W. Then, the following statements are equivalent:

1. C is listable on W;

2. there is a computable function $f : V_1^{n_1} \rightharpoonup W^m$ such that range $f = C$;

3. $\chi_{C:W}^P$ is computable;

4. there is a computable function $g : W^m \rightharpoonup V_2^{n_2}$ such that dom $g = C$.

Proof. Since W is decidable then W is listable by Proposition 2.5. Hence, by Proposition 2.4, W^m is listable. Let h_{W^m} be a computable enumeration of W^m. Moreover, since $n_1 > 0$ and V_1 is listable and infinite, so is $V_1^{n_1}$ (using again Proposition 2.4). Taking into account Proposition 2.9, let $h_{V_1^{n_1}}$ be a computable injective enumeration of $V_1^{n_1}$. Finally, since V_2 is assumed to be non-empty, let u be an element of $V_2^{n_2}$.

$(1 \to 4)$:

Given that C is listable on W, we have two cases to consider. If C is empty then it is the domain of the undefined function (that is, undefined everywhere) from W^m to $V_2^{n_2}$ and, so, (4) holds. Otherwise, take a computable enumeration h_C of C over W and let

$$g' = \lambda w . (\mu k . h_C(k) = w) : W^m \rightharpoonup \mathbb{N}.$$

Thus, $g'(w)$ is defined iff there is some k such that $h_C(k) = w$, that is, iff $w \in C$. Hence,

$$\operatorname{dom} g' = C.$$

Moreover, g' is a computable function by Proposition 1.6. Take

$$g = (\lambda k . u) \circ g' : W^m \rightharpoonup V_2^{n_2}.$$

Clearly, dom $g = \operatorname{dom} g' = C$. Moreover, by Proposition 1.2, g is computable. Thus, statement (4) holds.

$(4 \to 3)$:

Let $C = \operatorname{dom} g$ for some computable function $g : W^m \rightharpoonup V_2^{n_2}$. Let

$$g' = (\lambda v . 1) \circ g : W^m \rightharpoonup \mathbb{N}.$$

Clearly, g' is computable and

$$\operatorname{dom} g' = \operatorname{dom} g = C.$$

Furthermore,

$$g' = \lambda w . \begin{cases} 1 & \text{if } w \in C \\ \text{undefined} & \text{otherwise} \end{cases} = \chi_{C:w}^{P}$$

and, so, (3) holds.

$(3 \rightarrow 2)$:

Assume that $\chi_{C:w}^{P} : W^m \rightharpoonup \mathbb{N}$ is computable. Let

$$f' : W^m \rightharpoonup W^m$$

be the function computed by the following program:

```
function (w) (
    if P_{χ^P_{C:w}} (w) == 1 then
        return w
    else
        while true do null
)
```

where $P_{\chi_{C:w}^{P}}$ is a program that computes $\chi_{C:w}^{P}$. Observe that range $f' = \operatorname{dom} f' = \operatorname{dom} \chi_{C:w}^{P} = C$. Since, by hypothesis, $V_1^{n_1}$ is an infinite listable set on V_1 then, by Proposition 2.9, there is an injective computable enumeration of $V_1^{n_1}$ over V_1. Let $h_{V_1^{n_1}} : \mathbb{N} \to V_1^{n_1}$ be such an enumeration. Therefore, by Proposition 1.7, $(h_{V_1^{n_1}})^{-1}$ is also a computable map. Let

$$f = f' \circ h_{W^m} \circ (h_{V_1^{n_1}})^{-1} : V_1^{n_1} \rightharpoonup W^m.$$

Observe that f is a computable function by Proposition 1.2. It remains to prove that

$$\operatorname{range} f = C$$

and, so, statement (2) holds. Indeed,

(a) range $f \subseteq C$. Assume that $w \in \operatorname{range} f$. Then, there is v such that $f(v) = w$. That is,

$$f' \circ h_{W^m} \circ (h_{V_1^{n_1}})^{-1}(v) = w.$$

Therefore,

$$h_{W^m} \circ (h_{V_1^{n_1}})^{-1}(v) \in \operatorname{dom} f' = C$$

and so $w \in C$ since $w = h_{W^m} \circ (h_{V_1^{n_1}})^{-1}(v)$.

(b) $C \subseteq \mathrm{range}\, f$. Let $c \in C$. We must show that there is $v \in V_1^{n_1}$ such that

$$h_{W^m} \circ (h_{V_1^{n_1}})^{-1}(v) = c.$$

Since $c \in W^m$ let $k \in \mathbb{N}$ be such that $h_{W^m}(k) = c$. Let v be

$$h_{V_1^{n_1}}(k).$$

Then, $(h_{W^m}((h_{V_1^{n_1}})^{-1}(v))) = c$, thus, $f(v) = c$ and so $c \in \mathrm{range}\, f$.

$(2 \rightarrow 1)$:

Let $f : V_1^{n_1} \rightharpoonup W^m$ be computable with range $f = C$. Therefore, since $V_1^{n_1}$ is listable on V_1, $f(V_1^{n_1} \cap \mathrm{dom}\, f) = \mathrm{range}\, f = C$ is listable on W (by Proposition 2.7) and, thus, statement (1) holds. $\square$

   The following exercise extends some of the criteria in Proposition 2.11, within W, to other computable functions.

Exercise 2.5. Let $C \subseteq W^m$ be a set. Then:

   1. C is listable if and only if $C = \mathrm{range}\, f$ for some computable function $f : C_1 \rightharpoonup C_2$ where C_1 is an infinite listable set and C_2 is a decidable set;

   2. C is listable if and only if $C = \mathrm{dom}\, g$ for some computable function $g : D_1 \rightharpoonup D_2$ where D_1 and D_2 are decidable sets and D_2 is non-empty.

   The following result provides yet another very useful characterization of listable sets.

Proposition 2.12 (Projection Theorem). Let W be a working universe, $m, n \in \mathbb{N}^+$ and C a set of type m over W. Then

   1. If W is listable and C is listable on W then there exists a set R of type $m + n$ over W decidable on W such that
   $$C = R[1, m];$$

   2. If W is decidable and there exists a set R of type $m + n$ over W listable on W such that
   $$C = R[1, m]$$

   then C is listable on W.

Proof. Observe that condition $C = R[1, m]$ is equivalent to

$$(c_1, \ldots, c_m) \in C \quad \text{iff} \quad \exists (d_1, \ldots, d_n) \in W^n : (c_1, \ldots, c_m, d_1, \ldots, d_n) \in R.$$

(1) Let h_{W^n} be a computable injective enumeration of W^n. Such an enumeration exists by Proposition 2.9, thanks to the assumption that W is a listable working universe. If $C = \emptyset$, then take R to be the empty set of type $m + n$, which is clearly decidable on W and satisfies the required condition. Otherwise, suppose that C is enumerated by the computable map h_C. Take

$$R = \{(c_1, \ldots, c_m, d_1, \ldots, d_n) : k \in \mathbb{N}, \, h_C(k) = (c_1, \ldots, c_m), \, h_{W^n}(k) = (d_1, \ldots, d_n)\}$$

which is of the required type and obviously satisfies

$$(c_1, \ldots, c_m) \in C \quad \text{iff} \quad \exists (d_1, \ldots, d_n) \in W^n : (c_1, \ldots, c_m, d_1, \ldots, d_n) \in R.$$

It remains to verify that R is decidable on W. Let P_{h_C} and $\mathsf{P}_{(h_{W^n})^{-1}}$ be programs that compute h_C and $(h_{W^n})^{-1}$, respectively (recall that $(h_{W^n})^{-1}$ is a computable map by Proposition 1.7). Then the following algorithm

```
function (v₁,...,vₘ,w₁,...,wₙ) (
    x₁,...,xₘ = P_hC(P_(hWn)⁻¹(w₁,...,wₙ));
    if x₁ == v₁ ∧ ··· ∧ xₘ == vₘ then
        return 1
    else
        return 0
)
```

computes $\chi_{R:W}$.

(2) Suppose that R is a listable set on W of type $m + n$ and C is such that $C = R[1, m]$. Hence,

$$C = \mathrm{proj}_{[1,m]}^{W^{m+n}}(R).$$

Given that W is decidable then W^m is also decidable by Exercise 2.2. Therefore, since $\mathrm{proj}_{[1,m]}^{W^{m+n}} : W^{m+n} \to W^m$ is computable, C is also listable on W by Proposition 2.7. $\quad\square$

Remark 2.1. Given a function $f : C_1 \rightharpoonup C_2$, recall that

$$\mathrm{graph}\, f = \{(x, f(x)) : x \in \mathrm{dom}\, f\} \subseteq C_1 \times C_2.$$

The following result provides a necessary and sufficient condition for a function to be computable.

Proposition 2.13 (Graph Theorem). Let C_1 and C_2 be decidable sets with C_2 a nonempty set. Then, a function $f : C_1 \rightharpoonup C_2$ is computable if and only if $\mathrm{graph}\, f$ is listable.

Proof. ($\rightarrow$) Observe that, by Exercise 2.5, $\mathrm{dom}\, f$ is listable. Consider the function

$$g = \lambda\, w \,.\, (w, f(w)) : C_1 \rightharpoonup C_1 \times C_2.$$

Observe that g is

$$g = \langle id_{C_1}, f \rangle.$$

So g is computable because it is an aggregation of the computable map id_{C_1} and the computable function f. Clearly, graph $f = g(\operatorname{dom} f)$. Therefore, by Proposition 2.7, graph f is listable.

($\leftarrow$) If graph f is empty then f is undefined everywhere and, thus, computable. Otherwise, let $h : \mathbb{N} \to C_1 \times C_2$ be a computable enumeration of graph f. Observe that

$$\operatorname{proj}_{[1]}^{C_1 \times C_2} \circ h : \mathbb{N} \to C_1$$

is such that

$$\operatorname{range}\left(\operatorname{proj}_{[1]}^{C_1 \times C_2} \circ h\right) = \operatorname{dom} f.$$

Note also that

$$\min_{\operatorname{proj}_{[1]}^{C_1 \times C_2} \circ h}^{\leq_{\mathbb{N}}} : C_1 \to \mathbb{N}.$$

We are going to prove that

$$f = \operatorname{proj}_{[2]}^{C_1 \times C_2} \circ h \circ \min_{\operatorname{proj}_{[1]}^{C_1 \times C_2} \circ h}^{\leq_{\mathbb{N}}} : C_1 \rightharpoonup C_2.$$

Let $c_1 \in C_1$. Thus, we have one of the following cases:

(a) $c_1 \in \operatorname{dom} f$. Then, $\min_{\operatorname{proj}_{[1]}^{C_1 \times C_2} \circ h}^{\leq_{\mathbb{N}}}(c_1)$ is defined and so

$$c_1 \in \operatorname{dom} \operatorname{proj}_{[2]}^{C_1 \times C_2} \circ h \circ \min_{\operatorname{proj}_{[1]}^{C_1 \times C_2} \circ h}^{\leq_{\mathbb{N}}}.$$

Therefore:

$$\left(\operatorname{proj}_{[2]}^{C_1 \times C_2} \circ h \circ \min_{\operatorname{proj}_{[1]}^{C_1 \times C_2} \circ h}^{\leq_{\mathbb{N}}}\right)(c_1) = \left(\operatorname{proj}_{[2]}^{C_1 \times C_2} \circ h\right)(k)$$

$$= f(c_1)$$

where k is the least natural number such that $\left(\operatorname{proj}_{[1]}^{C_1 \times C_2} \circ h\right)(k) = c_1$.

(b) $c_1 \notin \operatorname{dom} f$. Therefore, $\min_{\operatorname{proj}_{[1]}^{C_1 \times C_2} \circ h}^{\leq_{\mathbb{N}}}(c_1)$ is not defined and so

$$\left(\operatorname{proj}_{[2]}^{C_1 \times C_2} \circ h \circ \min_{\operatorname{proj}_{[1]}^{C_1 \times C_2} \circ h}^{\leq_{\mathbb{N}}}\right)(c_1)$$

is not defined. Moreover, $f(c_1)$ is also not defined.

So, f is a computable function by Proposition 1.6 and Proposition 1.2. $\square$

2.3 Problem Reduction

In this section we concentrate our attention on decision problems. Moreover, we also analyze transference of (positive and negative) decidability and semidecidability results between decision problems.

Definition 2.3. A *(decision) problem* is a triple (W,n,C) where W is a working universe, $n \in \mathbb{N}^+$ and C a set of type n over W.

Remark 2.2. Traditionally a decision problem (W,n,C) is presented in the form of a question:

$$\text{Given } w \in W^n, \text{ does } w \in C?$$

Example 2.3. Consider the problem:

$$\text{Given } n \in \mathbb{N}, \text{ does } n \in \{n \in \mathbb{N} : n \text{ is a multiple of } 10\}?$$

As in many cases, this problem can be presented as follows:

$$\text{Given } n \in \mathbb{N}, \text{ is } n \text{ a multiple of } 10?$$

Example 2.4. Recall Example 1.3. The problem:

$$\text{Given a propositional formula } \alpha \in L_P, \text{ is } \alpha \text{ satisfiable?}$$

is well known in propositional logic. The interested reader can see the details in Section 9.2.

More examples are discussed in detail in Chapter 9.

Definition 2.4. The decision problem (W,n,C) is said to be *decidable* if C is decidable on W, that is, $\chi_{C:W} : W^n \to \mathbb{N}$ is computable. Moreover, the problem is said to be *semidecidable* if C is listable on W, that is, $\chi_{C:W}^P : W^n \to \mathbb{N}$ is computable.

When (W,n,C) is decidable we say that an algorithm that computes $\chi_{C:W}$ decides or solves the problem. Similarly, when (W,n,C) is semidecidable, we say that a procedure that computes $\chi_{C:W}^P$ semidecides or semisolves the problem.

Sometimes we can take advantage of knowing that a decision problem is decidable in order to conclude that another decision problem is also decidable. Similarly for undecidability and semidecidability. More precisely, given $C_1 \subseteq W_1^{n_1}$ and $C_2 \subseteq W_2^{n_2}$, we say that problem

$$\text{Given } w \in W_1^{n_1}, \text{ does } w \in C_1?$$

can be *reduced* to problem

$$\text{Given } w \in W_2^{n_2}, \text{ does } w \in C_2?$$

if any algorithm for deciding the second problem can be adapted to decide the first problem. In such a scenario, the decidability of C_2 entails the decidability of C_1. Conversely, the undecidability of C_1 entails the undecidability of C_2.

Definition 2.5. Problem (W_1, n_1, C_1) is *reducible* to problem (W_2, n_2, C_2) if there is a computable map $s : W_1^{n_1} \to W_2^{n_2}$ such that

$$w \in C_1 \text{ iff } s(w) \in C_2$$

for every $w \in W_1^{n_1}$.

When W_1 and W_2 are $\mathbb{N}$ and $n_1 = n_2 = 1$ this notion is called *many to one reducibility* and denoted by $\leq_{\mathrm{m}}$. We will discuss this relation in detail in Section 5.1.

Proposition 2.14. Assume that problem (W_1, n_1, C_1) is reducible to (W_2, n_2, C_2) via $s : W_1^{n_1} \to W_2^{n_2}$ and that W_2 is decidable. Then,

- If (W_2, n_2, C_2) is decidable then (W_1, n_1, C_1) is decidable;

- If (W_2, n_2, C_2) is semidecidable then (W_1, n_1, C_1) is semidecidable.

Proof. Assume that C_2 is decidable on W_2. Then, $\chi_{C_2:W_2}$ is a computable map. Note that $\chi_{C_1:W_1} = \chi_{C_2:W_2} \circ s$ taking into account that, by hypothesis, s reduces (W_1, n_1, C_1) to (W_2, n_2, C_2). So by Proposition 1.2 we conclude that $\chi_{C_1:W_1}$ is a computable map. Hence, C_1 is decidable on W_1. The second statement follows in a similar way. $\square$

Example 2.5. The problem

Given $n \in \mathbb{N}$, is n a multiple of 4?

is reducible to the problem

Given $n \in \mathbb{N}$, is n a multiple of 8?

via the computable map $s = \lambda x . 2x$ since

n is a multiple of 4   iff $2n$ is a multiple of 8.

Thus, by Proposition 2.14, the first problem is decidable since, by Example 2.1, the second problem is decidable.

2.4   Gödelization

In his proof of incompleteness of arithmetic, Kurt Gödel found that it was necessary to represent formulas and derivations of the calculus within the language of arithmetic, for which purpose he encoded formulas and derivations by natural numbers. In tribute to his work, such encodings are nowadays known as Gödelizations.

The idea in this section is to show that every question on a working universe W can be answered by a similar question over the working universe $\mathbb{N}$. With this purpose in mind, we start by introducing a map from W to $\mathbb{N}$ satisfying some properties that will ensure our goals.

Definition 2.6. Let W be a working universe. A map

$$g : W \to \mathbb{N}$$

is said to be a *Gödelization* of W if:

1. g is computable and injective;

2. $g(W)$ is a decidable set over $\mathbb{N}$;

3. $g^{-1} : g(W) \to W$ is a computable map.

Example 2.6. Consider propositional logic (Example 1.3) over $P = \{p_0, p_1, \dots\}$. We provide an example of a Gödelization of W_{L_P} inspired by [43]. Let

$$A = \{\langle, \rangle, "not", "implies"\} \cup \{\langle "p", k \rangle : k \in \mathbb{N}\}$$

and $h : A \to \mathbb{N}$ be such that

$$h = \lambda a . \begin{cases} 3 & \text{if } a \text{ is } \langle \\ 5 & \text{if } a \text{ is } \rangle \\ 7 & \text{if } a \text{ is "not"} \\ 9 & \text{if } a \text{ is "implies"} \\ 5 + 6(k+1) & \text{if } a \text{ is } \langle "p", k \rangle, k \in \mathbb{N} \end{cases} .$$

Then, the map

$$g = \lambda \alpha . p_0^{h((\alpha)_1)} \times \cdots \times p_{|\alpha|_A - 1}^{h((\alpha)_{|\alpha|_A})} : W_{L_P} \to \mathbb{N},$$

where $\{p_i\}_{i \in \mathbb{N}}$ stands for the sequence of prime numbers, $(\alpha)_k$ is the k-th symbol of A in α and $|\alpha|_A$ is the number of symbols of A in α, is a Gödelization of W_{L_P}. Indeed:

(1) g is a computable injective map. We leave to the interested reader to show that g is a computable map. With respect to the injectivity of g assume, by contradiction that $n = g(\alpha) = g(\alpha')$ and $\alpha \neq \alpha'$. Then, n would have two distinct prime factorizations which contradicts the Fundamental Theorem of Arithmetic [37].

(2) $g(W_{L_P})$ is decidable on $\mathbb{N}$. Then,

$$\chi_{g(W_{L_P}):\mathbb{N}} = \lambda n . \begin{cases} 1 & \text{if } n = 3, 5, 7, 9 \text{ or } n \text{ is of the form } 5 + 6(k+1), k \in \mathbb{N} \\ 0 & \text{otherwise} \end{cases}$$

is clearly a computable map.

(3) $g^{-1} : g(W_{L_P}) \to W_{L_P}$ is a computable map. We leave to the interested reader to work out the details of a program that computes this function.

Remark 2.3. Given a Gödelization g of a working universe W and $n \in \mathbb{N}^+$, we denote by

$$g^{(n)} : W^n \to \mathbb{N}^n$$

the map $\lambda\, w_1, \ldots, w_n \cdot (g(w_1), \ldots, g(w_n))$.

Exercise 2.6. Given a Gödelization g of a working universe W and $n \in \mathbb{N}^+$, show that $g^{(n)} : W^n \to \mathbb{N}^n$ and $(g^{(n)})^{-1} : g^{(n)}(W^n) \to W^n$ are injective and computable maps.

Proposition 2.15 (First Gödelization Theorem). Let g be a Gödelization of W. Then:

1. (W, n, C) is reducible to $(\mathbb{N}, n, g^{(n)}(C))$;

2. $(\mathbb{N}, n, g^{(n)}(C))$ is reducible to (W, n, C) whenever if $C = W^n$ then $\mathbb{N}^n = g^{(n)}(C)$.

Proof. (1) (W, n, C) reduces to $(\mathbb{N}, n, g^{(n)}(C))$. It is enough to observe that $g^{(n)} : W^n \to \mathbb{N}^n$ is computable by Exercise 2.6 and $w \in C$ iff $g^{(n)}(w) \in g^{(n)}(C)$ for each $w \in W^n$, since $g^{(n)}$ is injective by Exercise 2.6.

(2) $(\mathbb{N}, n, g^{(n)}(C))$ reduces to (W, n, C). Assume that if $C = W^n$ then $\mathbb{N}^n = g^{(n)}(C)$. We consider two cases:

(a) $C \neq W^n$. Let $w \in W^n \setminus C$. Consider the map $s : \mathbb{N}^n \to W^n$ such that

$$s = \lambda\, x_1, \ldots, x_n \cdot \begin{cases} (g^{(n)})^{-1}(x_1, \ldots, x_n) & \text{if } (x_1, \ldots, x_n) \in g^{(n)}(W^n) \\ w & \text{otherwise.} \end{cases}$$

Then s is well defined by Exercise 2.6, and is clearly computable since $g(W)$ is decidable on $\mathbb{N}$ and $(g^{(n)})^{-1}$ is computable by Exercise 2.6. Finally, $(x_1, \ldots, x_n) \in g^{(n)}(C)$ iff $s(x_1, \ldots, x_n) \in C$. The implication from left to right is immediate. For the other direction, let $s(x_1, \ldots, x_n) \in C$. Since $w \notin C$ then $(x_1, \ldots, x_n) \in g^{(n)}(W^n)$ and

$$s(x_1, \ldots, x_n) = (g^{(n)})^{-1}(x_1, \ldots, x_n) \in C.$$

Thus,

$$g^{(n)}((g^{(n)})^{-1}(x_1, \ldots, x_n)) = (x_1, \ldots, x_n) \in g^{(n)}(C).$$

(b) $C = W^n$. Then $\mathbb{N}^n = g^{(n)}(C)$. Let $s : \mathbb{N}^n \to W^n$ be such that

$$s = (g^{(n)})^{-1}.$$

Therefore s is computable because $(g^{(n)})^{-1}$ is computable by Exercise 2.6. Moreover, $(x_1, \ldots, x_n) \in \mathbb{N}^n$ iff $s(x_1, \ldots, x_n) \in W^n$. $\qquad\square$

The next result follows immediately by Proposition 2.15 and Proposition 2.14.

Proposition 2.16. Let (W, n, C) be a problem and g a Gödelization of W such that if $C = W^n$ then $\mathbb{N} = g^{(n)}(C)$. Then,

- (W, n, C) is decidable iff $(\mathbb{N}, n, g^{(n)}(C))$ is decidable;

- (W, n, C) is semidecidable iff $(\mathbb{N}, n, g^{(n)}(C))$ is semidecidable.

The next result shows that computability over any working universe can be discussed simply over the working universe of natural numbers.

Proposition 2.17 (Second Gödelization Theorem). Let g_1 and g_2 be Gödelizations of W_1 and W_2, respectively, and C_1 and C_2 sets of types n_1 and n_2 over W_1 and W_2, respectively. Assume that $g_1^{(n_1)}(C_1)$ and $g_2^{(n_2)}(C_2)$ are decidable. Then:

$$f : C_1 \rightharpoonup C_2 \text{ is computable}$$

if and only if

$$\left(g_2^{(n_2)}|_{C_2}\right)^\bullet \circ f \circ \left(\left(g_1^{(n_1)}\right)^{-1}\Big|_{g_1^{(n_1)}(C_1)}\right)^\bullet : g_1^{(n_1)}(C_1) \rightharpoonup g_2^{(n_2)}(C_2) \text{ is computable.}$$

Proof. The proof follows immediately by Proposition 1.1, Exercise 1.4, Proposition 1.2 and Proposition 2.2 since g_1 and g_2 are injective and computable maps. □

Accordingly, from now on, unless otherwise stated, by a set of type n we mean a subset of $\mathbb{N}^n$ and by a function/map we mean a function/map with arguments and values in $\mathbb{N}$. Moreover, by computable function we mean computable over $\mathbb{N}$, by decidable set we mean decidable set over $\mathbb{N}$ and by listable we mean listable over $\mathbb{N}$.

Remark 2.4. The following notation will be useful:

- $\mathscr{S} = \bigcup_{n \in \mathbb{N}^+} \mathscr{S}_n$ where $\mathscr{S}_n$ is the class of sets of type n over $\mathbb{N}$ for each $n \in \mathbb{N}^+$;

- $\mathscr{D} = \bigcup_{n \in \mathbb{N}^+} \mathscr{D}_n$ where $\mathscr{D}_n$ is the set of decidable sets in $\mathscr{S}_n$ for each $n \in \mathbb{N}^+$;

- $\mathscr{L} = \bigcup_{n \in \mathbb{N}^+} \mathscr{L}_n$ where $\mathscr{L}_n$ is the set of listable sets in $\mathscr{S}_n$ for each $n \in \mathbb{N}^+$;

- $\mathscr{F} = \bigcup_{n \in \mathbb{N}, m \in \mathbb{N}^+} \mathscr{F}_n^m$ where $\mathscr{F}_n^m$ is the class of functions from $\mathbb{N}^n$ to $\mathbb{N}^m$ for each $n \in \mathbb{N}$ and $m \in \mathbb{N}^+$;

- $\mathscr{C} = \bigcup\limits_{n\in\mathbb{N},m\in\mathbb{N}^+} \mathscr{C}_n^m$ where $\mathscr{C}_n^m$ is the class of computable functions in $\mathscr{F}_n^m$ for
  each $n \in \mathbb{N}$ and $m \in \mathbb{N}^+$;

- $\mathscr{F}_n = \mathscr{F}_n^1$ for each $n \in \mathbb{N}$;

- $\mathscr{C}_n = \mathscr{C}_n^1$ for each $n \in \mathbb{N}$;

- $\mathrm{m}\mathscr{F}$, $\mathrm{m}\mathscr{F}_n^m$ and $\mathrm{m}\mathscr{F}_n$ are the classes of maps in $\mathscr{F}$, $\mathscr{F}_n^m$ and $\mathscr{F}_n$, respectively;

- $\mathrm{m}\mathscr{C}$, $\mathrm{m}\mathscr{C}_n^m$ and $\mathrm{m}\mathscr{C}_n$ are the classes of maps in $\mathscr{C}$, $\mathscr{C}_n^m$ and $\mathscr{C}_n$, respectively.

The Gödelization theorems explain why Computability Theory has traditionally been developed in the realm of natural numbers using only functions having powers of $\mathbb{N}$ for source and target sets. Following that approach, when one needs to study computability within another domain of objects (for example, in that of the formulas of first-order logic or in that of the programs) one begins by intuitively introducing a Gödelization and afterward work only with their Gödel numbers.

Clearly, in the traditional setting it is not possible to prove results for Gödelizations themselves, since they are outside the scope of that simplified theory and, so, the Gödelization theorems are accepted as postulates. In fact, the need for proving the Gödelization theorems was the major reason for adopting in this course a more general setting, encompassing different computation realms, albeit all contained in W.

The results about Gödelizations fully justify that we transfer our interest from the realm of words in W to the realm of natural numbers. However, it will still be necessary to go back to the more general setting in a few cases.

2.5   Solved Problems and Exercises

Problem 2.1. Show that $\chi_{C:W}^{\mathrm{p}}$ is a computable function whenever C is listable on W and W is decidable (do not use the Listability Criteria, Proposition 2.11).

Solution Let C be a set of type n listable on W. Consider two cases:

(1) $C = \emptyset$. Then, $\chi_{C:W}^{\mathrm{p}} = (\lambda x.\,\mathrm{undefined})$ is a computable function (see Example 1.16).

(2) $C \neq \emptyset$. Then, there is a computable enumeration $h : \mathbb{N} \to W^n$ of C. We now show that

$$\chi_{C:W}^{\mathrm{p}} = (\lambda x.\,1) \circ \min\nolimits_h^{\leq\mathbb{N}} : W^n \rightharpoonup \mathbb{N}$$

and so is computable since $\min_h^{\leq\mathbb{N}}$ is computable (by Proposition 1.6) and since the composition of computable functions is computable (Proposition 1.2). It is enough to show that

$$C = \mathrm{dom}\left((\lambda x.\,1) \circ \min\nolimits_h^{\leq\mathbb{N}}\right)$$

since both functions have 1 as the result for each element in the domain. Indeed:

($\subseteq$) Let $w \in C$. Then, there is a natural number whose image by h is w. Hence, $\min_h^{\leq \mathbb{N}}(w)$ is defined. Therefore, $((\lambda x.1) \circ \min_h^{\leq \mathbb{N}})(w)$ is defined and as a consequence $w \in \mathrm{dom}\,((\lambda x.1) \circ \min_h^{\leq \mathbb{N}})$.

($\supseteq$) Take $w \in \mathrm{dom}\,((\lambda x.1) \circ \min_h^{\leq \mathbb{N}})$. Then, $w \in \mathrm{dom}\min_h^{\leq \mathbb{N}}$. Hence, $\min_h^{\leq \mathbb{N}}(w)$ is defined and belongs to $\mathbb{N}$. Therefore, there is a natural number whose image by h is w. Thus, $w \in \mathrm{range}\,h = C$. ∇

Problem 2.2. Show that if C is a set over W with computable characteristic function and W is listable then C is listable.

Solution Assume that W is listable and that C is a set of type n over W such that the characteristic function of C

$$\chi_{C:W}^{p} = \lambda w. \begin{cases} 1 & \text{if } w \in C \\ \text{undefined} & \text{otherwise} \end{cases} : W^n \rightharpoonup \mathbb{N}$$

is computable. Let $\mathrm{P}_{\chi_{C:W}^{p}}$ be a program that computes $\chi_{C:W}^{p}$. There are two cases to consider:

(1) $C = \emptyset$. Then, C is listable.

(2) $C \neq \emptyset$. Let $c \in C$. Observe that W^n is listable because W is listable (Proposition 2.4). Let

$$h : \mathbb{N} \to W^n$$

be a computable enumeration of W^n, computed by P_h. We want to find a computable enumeration of C. Consider the following program P:

```
function (k) (
    k₁ = P_K(k);
    k₂ = P_L(k);
    if isnat(stceval(P_{χ_{C:W}^p}, P_h, k₁, k₂, −1)) then
        return P_h(k₁)
    else
        return c
)
```

Let h_C be the function computed by P, that is, the function that given $k \in \mathbb{N}$ is such that

$$h_C(k) = \begin{cases} x & \text{if the execution of } \mathrm{P} \text{ on } k \text{ returns } x \\ \text{undefined} & \text{otherwise} \end{cases}.$$

We now show that h_C is a computable enumeration of C. Indeed:

(a) h_C is a map since the execution of P on any input always terminates (the evaluation of $\mathrm{stceval}(P_{\chi_{C:W}^P}, P_h, k_1, k_2, -1)$ on any k_1 takes at most k_2 units of time).

(b) $h_C : \mathbb{N} \to W^n$. In fact, P either returns $c \in C$ or returns $h(k_1) \in C$.

(c) h_C is an onto map. Let $d \in C$. Then, there exists $k_1 \in \mathbb{N}$ such that $h(k_1) = d$ since h is an enumeration of W^n. Assume that the execution of $P_{\chi_{C:W}^P}$ on d terminates in k_2 units of time with result 1. Then, the execution of P on $J(k_1, k_2)$ also terminates in a finite amount of time returning $h(k_1)$ which is d. ∇

Problem 2.3. Let $f, g : \mathbb{N} \to \mathbb{N}$ be computable maps. Furthermore, assume that if $j \in \mathrm{range}\, f$ then there is $k \in \mathbb{N}$ such that $k \le g(j)$ and $j = f(k)$. Show that $\mathrm{range}\, f$ is decidable on $\mathbb{N}$.

Solution Given $j \in \mathbb{N}$ the goal is to decide if $j \in \mathrm{range}\, f$, that is, if there is $k \in \mathbb{N}$ such that $j = f(k)$. Based on the conditions for f and g, we can restrict the search for k up to $g(j)$. Consider the following program P:

```
function ( j ) (
    i = 0;
    r = 0;
    while i ≤ Pg(j) ∧ r == 0 do (
        if Pf(i) == j then
            r = 1;
        i = i+1
    );
    return r
)
```

where P_f and P_g are programs that compute f and g, respectively. Clearly, P is an algorithm as the loop terminates for every value of $j \in \mathbb{N}$. It remains to check that P computes

$$\chi_{\mathrm{range}\, f:\mathbb{N}} : \mathbb{N} \to \mathbb{N}.$$

Indeed:

(1) $j \in \mathrm{range}\, f$. Then, by the hypothesis, there is $k \in \mathbb{N}$ such that $k \le g(j)$ and $f(k) = j$. When i reaches k, then r is set to 1, and the execution of P terminates returning 1.

(2) Otherwise, there is no $k \in \mathbb{N}$ such that $f(k) = j$. Hence, the loop will terminate when i reaches $g(j)$ and the execution of P returns 0. ∇

Problem 2.4. Show that $C \subseteq \mathbb{N}$ is infinite and decidable on $\mathbb{N}$ if and only if there is a computable map $f : \mathbb{N} \to \mathbb{N}$ such that:

- $C = \mathrm{range}\, f$;

- f is a strictly increasing map.

Solution ($\rightarrow$) Assume that $C \subseteq \mathbb{N}$ is infinite and decidable on $\mathbb{N}$. Consider the following program P:

```
function (k) (
    j = 0;
    i = 0;
    while i ≤ k do (
        if P_{χ_{C:N}}(j) == 1 then (
            i = i + 1;
            r = j
        );
        j = j + 1
    );
    return r
)
```

where $P_{\chi_{C:\mathbb{N}}}$ is a program that computes $\chi_{C:\mathbb{N}}$. Let $f : \mathbb{N} \to \mathbb{N}$ be the function computed by P. Then:

(1) For each $k \in \mathbb{N}$, the execution of P on k terminates with the $(k+1)$-th minimum element of C as the result, as we now prove by induction on k.

(Base) $k = 0$. The execution of the loop terminates when the value of i is greater than 0. For this to happen, i must have been incremented which can only happen when $j \in C$. Since C is non-empty there is always such a j. Furthermore, assume that when i reaches 1, j has value m. This means that $0, \ldots, m-1 \notin C$ and that $m \in C$. Hence m is the first minimum element of C, which is the value returned by the execution of P on 0.

(Step) $k > 0$. The execution of P on k includes the steps of the execution of P on $k-1$ that assigns to r the k-th minimum of C. Note that at this point $i = k$. The loop will continue until i reaches $k+1$, which will happen the next time $\chi_{C:\mathbb{N}}(j) = 1$. Observe that there is such a j because C is an infinite set. In this case, the execution of P on k returns the $(k+1)$-th minimum element of C.

(2) f is an enumeration. Take $m \in C$. Then, m is the $k+1$-th minimum element of C for some $k \in \mathbb{N}$. Hence, the execution of P on k will terminate returning m. Thus, $f(k) = m$.

(3) f is a strictly increasing map. Take $k_1 < k_2$. Then, $f(k_1)$ is the $(k_1 + 1)$-th minimum element of C and $f(k_2)$ is the $(k_2 + 1)$-th minimum element of C. As the $(k_1 + 1)$-th minimum element is less than the $(k_2 + 1)$-th minimum element then $f(k_1) < f(k_2)$.

($\leftarrow$) To prove the converse, we start by observing that f is injective. As a consequence, range $f = C$ is infinite. To prove that C is decidable, we will use Problem 2.3, and we will find a computable map g satisfying the necessary requirements. Choose $g = \lambda x.x$. We start by proving that

$$i \leq f(i)$$

for every $i \in \mathbb{N}$ by induction on i.

(Base) $i = 0$. Straightforward since $0 \leq f(0)$.

(Step) Clearly $f(i) < f(i+1)$, by hypothesis. Furthermore, $i \leq f(i)$, by the induction hypothesis. Hence, $i < f(i+1)$ which implies that $i+1 \leq f(i+1)$.

Let $j \in$ range f. Then, there exists $k \in \mathbb{N}$ such that $f(k) = j$. Furthermore, we also have that $k \leq f(k) = j = g(j)$, that is, $k \leq g(j)$, as required. Hence, by Problem 2.3, we conclude that $C =$ range f is decidable on $\mathbb{N}$. $\triangledown$

Problem 2.5. Show that every infinite listable set of natural numbers contains an infinite decidable set.

Solution Let $D \subseteq \mathbb{N}$. If D is infinite and listable then there is a computable enumeration $h_D : \mathbb{N} \to \mathbb{N}$ of D. Consider the following program P:

```
function (k) (
    if k == 0 then
        return P_{h_D}(0)
    else (
        i = 0;
        while P_{h_D}(i) ≤ P_{h_D}(k − 1) do
            i = i + 1;
        return P_{h_D}(i)
    )
)
```

where P_{h_D} is a program that computes h_D. Let $f : \mathbb{N} \to \mathbb{N}$ be the map computed by P. We now show that f satisfies the requirements of Problem 2.4. Observe that:

(1) P is an algorithm. Indeed, the execution of P on $k \in \mathbb{N}$ will always terminate since the execution of P_{h_D} on any natural number will always terminate and because, for every natural number $m \in D$, there is a natural number $m' \in D$ such that $m < m'$ since D is infinite.

(2) f is a strictly increasing map. Indeed:

- $f(0)$ is $h_D(0)$;

- $f(k+1)$ is the next element in the enumeration h_D of D greater than $f(k)$.

Therefore, by Problem 2.4, range f is infinite and decidable. The thesis follows since range $f \subseteq D$. Indeed, the execution of P always terminates returning $h_D(i)$ for some $i \in \mathbb{N}$. $\triangledown$

Supplementary Exercises

Exercise 2.7. Recall Example 1.3. Show that W_{L_P} is decidable.

Exercise 2.8. Can the hypotheses in Proposition 2.11 be weakened when we are interested in proving only one of the implications (e.g. $1 \rightarrow 3$)?

Exercise 2.9. Let $C_1 \in \mathcal{L}$ and $f : C_1 \rightharpoonup C_2 \in \mathcal{C}$. Show that $\operatorname{dom} f \in \mathcal{L}$.

Exercise 2.10. Let $f, g \in \mathcal{C}_1$. Assume $f(\operatorname{dom} g) \neq \emptyset$. Present an algorithm that computes an enumeration of $f(\operatorname{dom} g)$.

Exercise 2.11. Show that every set decidable on W is decidable if W is decidable. Moreover, show that every decidable set is decidable on W.

Exercise 2.12. Show that every finite subset of W is decidable on W.

Exercise 2.13. Let $n \in \mathbb{N}^+$. Show that β_n^1 is a Gödelization of $\mathbb{N}^n$. What can be concluded about decidability and listability?

Exercise 2.14. Let $n \in \mathbb{N}^+$. Show that J_{W^n} is a Gödelization of W^n. What can be concluded about decidability and listability?

Chapter 3

Universality and Undecidability

The notion of universality was one of the major contributions of Alan Turing (see [63]) where he introduced the notion of universal machine. It led to the development of Computability Theory and paved the way for the advent of computers as we know them today. In fact, the first computers were called "universal computers" precisely for this reason. What distinguishes a universal computer is its capability of being programmable so that it can perform any computation, that is, it can compute any computable function.

Turing conceived a universal machine capable of emulating all machines of a kind proposed by him for capturing the notion of computable function. Such machines are nowadays known as Turing machines (see Chapter 7). Herein, we present universality within our chosen formalization of computable function (a function that can be computed by a program).

3.1 Universal Functions

Before defining the concept of universal function we introduce an auxiliary notion.

Definition 3.1. Given $u \in \mathscr{F}_2$ and $p \in \mathbb{N}$, the unary function

$$u_p = \lambda x . u(p,x)$$

is said to be the *section* of u at p.

Observe that $u(p,q) = u_p(q)$ for every $p,q \in \mathbb{N}$.

Exercise 3.1. Show that if $u \in \mathscr{C}_2$ then $u_p \in \mathscr{C}_1$ for every $p \in \mathbb{N}$. Moreover, show that every section of a binary map is a map and that every section of a function with finite domain also has finite domain.

Definition 3.2. A function $u \in \mathscr{F}_2$ is said to be *universal for* $C \subseteq \mathscr{F}_1$ if

$$C = \{u_p : p \in \mathbb{N}\}.$$

Moreover, u is said to be a *universal function* whenever u is universal for $\mathscr{C}_1$.

In other words, a binary function u is universal for a class C of unary functions if and only if every section of u is in C and vice-versa.

Exercise 3.2. Show that every binary function is universal for the class of its own sections.

Proposition 3.1. There is no function universal for $\mathscr{F}_1$.

Proof. We use a diagonal argument (diagonal arguments were introduced by Georg Cantor for proving that there are infinite sets that cannot be put in a one-to-one correspondence with the set of natural numbers). Assume, by contradiction that u is a universal function for $\mathscr{F}_1$. Let

$$f = \lambda x. \begin{cases} u_x(x) + 1 & x \in \mathrm{dom}\, u_x \\ 0 & \text{otherwise.} \end{cases}$$

It is clear that $f \in \mathscr{F}_1$ and so there is $p \in \mathbb{N}$ such that

$$(\dagger) \quad f = u_p.$$

Consider two cases:
(1) $p \notin \mathrm{dom}\, u_p$. Then, $f(p) \neq u_p(p)$ since $f(p) = 0$ and $u_p(p)$ is either undefined or is not in $\mathbb{N}$. Hence, $f \neq u_p$ contradicting ($\dagger$).
(2) $p \in \mathrm{dom}\, u_p$. Then,

$$f(p) = u_p(p) + 1 \neq u_p(p)$$

and so $f \neq u_p$ contradicting ($\dagger$).   □

Although there is no universal function for $\mathscr{F}_1$, we are able to find a universal function and, furthermore, a computable universal function as the following result states.

Definition 3.3. Let

$$\mathrm{univ} : \mathbb{N}^2 \rightharpoonup \mathbb{N}$$

be the function computed by the following program P_{univ}

```
function (p,x) (
    y = Pw(p)(x);
    if isnat(y) then
        return y
    else
        while true do null
)
```

where P_W is an algorithm that computes $\lambda k.k_W$ (recall Exercise 1.8). So

$$\text{univ} = \lambda\, p,x.\begin{cases} y & \text{if } y \in \mathbb{N} \text{ results from the execution of } p_W \text{ on } x \\ \text{undefined} & \text{otherwise} \end{cases}$$

Depending on the natural number p at hand, the result of the execution of P_W on p may not parse to a well formed program. In this case, univ_p is the unary function undefined everywhere. Otherwise, the program that results from the execution of P_W on p, when executed on $x \in \mathbb{N}$, may return a value outside of $\mathbb{N}$, in which case we make sure that the computation runs forever.

Proposition 3.2. Function univ is computable and universal with respect to $\mathscr{C}_1$.

Proof. Function univ is computable by definition. Regarding universality note that:

(1) $\{\text{univ}_p : p \in \mathbb{N}\} \subseteq \mathscr{C}_1$:

It is immediate to see that the program P_{univ_p}

$$\text{function } (x) \ (\text{return } P_{\text{univ}}(p,x))$$

computes univ_p for each $p \in \mathbb{N}$. So, $\text{univ}_p \in \mathscr{C}_1$ for each $p \in \mathbb{N}$.

(2) $\mathscr{C}_1 \subseteq \{\text{univ}_p : p \in \mathbb{N}\}$:

Let $f \in \mathscr{C}_1$ and P_f be a program that computes f. So, there is a unique $p \in \mathbb{N}$ such that $p_W = P_f$. Hence, the execution of P_W on p returns P_f. Let $x \in \mathbb{N}$. Consider two cases:

(a) Let $x \in \text{dom}\, f$. Then the execution of P_f, that is p_W, on x terminates in a finite number of steps returning $f(x)$. On the other hand, by definition, $\text{univ}(p,x)$ is the result of the execution of p_W on x. Therefore, $\text{univ}(p,x) = f(x)$.

(b) Assume that $x \notin \text{dom}\, f$. Observe that, the execution of P_{univ_p} on x includes the execution of p_W, that is P_f, on x that either does not terminate or it does with a result not in $\mathbb{N}$, triggering the execution of the infinite loop in P_{univ}. Thus, the execution of P_{univ_p} on x does not terminate and so $x \notin \text{dom}\,\text{univ}_p$.

Thus, we can conclude that $f = \text{univ}_p$. $\qquad\qquad\square$

Exercise 3.3. Propose the notion of universal function for $\mathscr{C}_n$ and show that such a function is computable.

At this point, it is natural to ask if there exists a computable map universal for $m\mathscr{C}_1$. Proposition 3.2 guarantees that there is a computable binary function whose sections include every computable map (among all computable functions). But, the non-existence of a computable binary map with sections encompassing all computable maps is easy to prove using a diagonal argument.

Proposition 3.3. There is no computable binary map h such that

$$m\mathscr{C}_1 \subseteq \{h_p : p \in \mathbb{N}\}.$$

Proof. Assume, by contradiction, that h is a computable binary map with $m\mathscr{C}_1 \subseteq \{h_p : p \in \mathbb{N}\}$. Then

$$e = \lambda k . h_k(k) + 1$$

is a computable unary map and so is in $m\mathscr{C}_1$. Observe that each section h_p differs at p from e. Therefore, $e \notin \{h_p : p \in \mathbb{N}\}$ and, so, $\{h_p : p \in \mathbb{N}\}$ does not contain every computable unary map contradicting our hypothesis. $\qquad\square$

Thus, given the key role of universality, there is no point in trying to develop the theory of computability only for maps. Partiality is essential.

Remark 3.1. Given $u \in \mathscr{F}_2$, we denote by

$$\text{diag}^u, \text{sdiag}^u : \mathbb{N} \rightharpoonup \mathbb{N}$$

the functions such that $\text{diag}^u = \lambda k . u(k,k)$ and $\text{sdiag}^u = \lambda k . u(k,k) + 1$.

The diagonal technique when applied to computable functions is useful for establishing the following two results.

Proposition 3.4. Let u be a computable universal function. Then, diag^u is a computable unary function such that no computable unary function differs from it everywhere.

Proof. It is immediate to see that diag^u is a computable unary function. On the other hand, for every computable unary function f there is $p \in \mathbb{N}$ such that $f = u_p$ and, so, $f(p) = u_p(p) = u(p,p) = \text{diag}^u(p)$. $\qquad\square$

Proposition 3.5. Let u be a computable universal function. Then, sdiag^u is computable and there is no computable unary map that extends sdiag^u.

Proof. Assume, by contradiction, that there is a computable unary map f extending sdiag^u. We show that f differs from diag^u everywhere, contradicting Proposition 3.4. Let $k \in \mathbb{N}$. Consider two cases.

(1) $k \in \text{dom diag}^u$. Since

$$\text{dom diag}^u = \text{dom sdiag}^u \subseteq \text{dom } f$$

then $f(k) \neq \text{diag}^u(k)$ because $f(k) = \text{diag}^u(k) + 1$.

(2) $k \notin \text{dom diag}^u$. Then, $k \notin \text{dom sdiag}^u$. On the other hand $k \in \text{dom } f$. So $f(k) \neq \text{diag}^u(k)$. □

Remark 3.2. Given a binary function u, we denote by

$$\mathbb{K}^u$$

the set $\text{dom diag}^u = \text{dom sdiag}^u$. Moreover, we denote by

$$\mathbb{K}$$

the set $\mathbb{K}^{\text{univ}}$.

Note that Proposition 3.5 entails that function sdiag^u is not a map when u is a computable universal function. So in this case,

$$\mathbb{K}^u \subsetneq \mathbb{N}.$$

Observe also that there are maps that extend sdiag^u which, by Proposition 3.5, are not computable.

3.2 Non-Decidable Listable Sets

With the results of the previous section at hand, we are finally ready to show that listability does not coincide with decidability.

Proposition 3.6. Let u be a computable universal function. Then,

$$\mathbb{K}^u$$

is listable but non-decidable.

Proof. Observe that, by Proposition 2.11, $\mathbb{K}^u$ is listable since it is the domain of the computable function sdiag^u. On the other hand, $\mathbb{K}^u$ is non-decidable as we now show. Assume, by contradiction, that $\mathbb{K}^u$ is decidable. Then, the map

$$f = \lambda x. \begin{cases} \text{sdiag}^u(x) & \text{if } x \in \mathbb{K}^u \\ 0 & \text{otherwise} \end{cases}$$

is computable and extends sdiag^u, in contradiction with Proposition 3.5. □

Proposition 3.7. Let u be a computable universal function. Then,

$$(\mathbb{K}^u)^c$$

is non-listable.

Proof. Observe that, by Proposition 3.6, $\mathbb{K}^u$ is a listable non-decidable set. So, by Post's Theorem, Proposition 2.6, $(\mathbb{K}^u)^c$ is non-listable since otherwise both would be decidable. □

As a consequence the function $\chi^p_{\mathbb{K}^c}$ is not computable, our first example of a non-computable function with non-listable domain.

In computing, a problem is dealt with by writing a program for solving it. Unfortunately, as we shall see in subsequent chapters, not every problem can be dealt with in this manner. The topic got much attention when the great mathematician David Hilbert in 1928 raised the problem "Is there a definite method or a *mechanical process* that can be applied to any mathematical assertion, and which is guaranteed to produce a correct decision as to whether that assertion is true?". The first examples that such a method does not exist in general were found independently by Kurt Gödel, the famous incompleteness theorems of arithmetic [20], Alonzo Church, first-order logic is non-decidable in general [9, 10] and Alan Turing, the nowadays famous halting problem [63]. Many other unsolvable problems have been identified since then. In fact, as the set of programs is denumerable, most problems are unsolvable.

We now show that the halting problem is non-decidable. We start with a variant of this problem.

Definition 3.4. The problem

$$(\mathbb{N}, 1, \{x \in \mathbb{N} : x \in \mathrm{dom}\,\mathrm{univ}_x\})$$

is called DHalting.

The question involved in the previous definition can be stated as follows:

Given $k \in \mathbb{N}$, does the execution of k_W on k returns a natural number?

Proposition 3.8. The decision problem DHalting is semidecidable but non-decidable.

Proof. Observe that $\mathbb{K} = \{x \in \mathbb{N} : x \in \mathrm{dom}\,\mathrm{univ}_x\}$. Since, by Proposition 3.6, $\mathbb{K}$ is listable, then DHalting is semidecidable. Moreover, $\mathbb{K}$ is non-decidable by the same result. Hence, DHalting is non-decidable. □

Definition 3.5. The problem

$$(\mathbb{N}, 2, \mathrm{dom}\,\mathrm{univ})$$

is called Halting.

The question involved in the previous definition can be stated as follows:

Given $p, x \in \mathbb{N}$, does the execution of p_W on x returns a natural number?

Proposition 3.9. The decision problem Halting is semidecidable but non-decidable.

Proof. (1) Halting is non-decidable.

Let $s = \lambda k.(k,k) : \mathbb{N} \to \mathbb{N}^2$. It is immediate to see that s is a computable map. Moreover, s reduces DHalting to Halting. Indeed

$$k \in \text{DHalting iff } (k,k) \in \text{Halting}.$$

Assume by contradiction that Halting is decidable. Hence, by Proposition 2.14, DHalting would also be decidable contradicting Proposition 3.8.

(2) Observe that $(p,x) \in \text{Halting}$ sse $(p,x) \in \text{dom univ}$, for every $(p,x) \in \mathbb{N}^2$. Since dom univ is a listable set then Halting is semidecidable. □

The halting problem[1] was first proved to be non-decidable by Alan Turing in 1936, at the time giving yet another negative answer to Hilbert's ambitious program of mechanizing mathematical reasoning (*"Entscheidungsproblem"*). For details, see [28, 29] and [42] for the Hilbert's Tenth Problem.

3.3 Proper Universal Functions

We now proceed to explore the properties of the enumeration of the class of computable functions that each computable universal function provides.

Remark 3.3. Given a binary function v, we denote by

$$\phi^v$$

the enumeration $\lambda p.v_p : \mathbb{N} \to \{v_p : p \in \mathbb{N}\}$. Moreover, we denote by

$$\phi_p^v$$

the function $\phi^v(p) = v_p$. Finally, we denote by

$$\phi$$

the enumeration $\phi^{\text{univ}} : \mathbb{N} \to \mathscr{C}_1$.

Definition 3.6. Given a universal function u and $f \in \mathscr{C}_1$, we say that p is a *u-index of f* whenever $f = u_p$.

---

[1] In a quite different formulation using tape automata now known as Turing machines.

Note that each computable unary function may have several u-indices. For instance, the set of univ-indices of each computable unary function is denumerable. The nature of these sets will be clarified later on, in Sections 4.1, 4.2 and 4.3, for the case of universal functions enjoying the following property:

Definition 3.7. A function $u \in \mathscr{C}_2$ is said to enjoy the *s-m-n property* if, for every $v \in \mathscr{C}_2$, there exists $s \in m\mathscr{C}_1$ such that

$$v_p = \phi^{u}_{s(p)}$$

for every $p \in \mathbb{N}$.

Intuitively, the translation map s provides the means for obtaining each section of v as a section of u (in an algorithmic way provided by s).

Definition 3.8. A *proper universal function* is a computable universal function enjoying the *s-m-n* property.

In fact, every computable binary function enjoying the *s-m-n* property is universal.

Proposition 3.10. Let $u \in \mathscr{C}_2$. Then u is universal whenever u enjoys the *s-m-n* property.

Proof. Assume that u has the *s-m-n* property. We have to check that

$$\mathscr{C}_1 = \{u_p : p \in \mathbb{N}\}.$$

(1) $\mathscr{C}_1 \subseteq \{u_p : p \in \mathbb{N}\}$.
Let $f \in \mathscr{C}_1$. Consider $v : \mathbb{N}^2 \rightharpoonup \mathbb{N}$ defined as follows:

$$v = \lambda\, p, x.\, f(x).$$

Clearly, function v is computable. Since u enjoys the *s-m-n* property, there is a computable map $s : \mathbb{N} \to \mathbb{N}$ such that

$$v(p, x) = u(s(p), x)$$

for every $p, x \in \mathbb{N}$. Hence,

$$f(x) = u(s(p), x) = u_{s(p)}(x)$$

for every $x \in \mathbb{N}$ and so $f \in \{u_p : p \in \mathbb{N}\}$.
(2) $\{u_p : p \in \mathbb{N}\} \subseteq \mathscr{C}_1$.
This fact follows from Exercise 3.1 since u is computable.   □

We will show in Section 4.1 that there are functions that are universal but not proper universal.

Before proceeding with the proof of the existence of proper universal functions, we need some notation. Recall maps J_W and $\lambda k . k_W$ defined in Section 1.4.

Remark 3.4. Given a program P, we denote by

$$\widehat{P}$$

the natural number $J_W(P)$.

In other words, $\widehat{P}$ is the code of program P. Clearly,

$$(\widehat{P})_W = P.$$

Indeed,

$$(\widehat{P})_W = (J_W(P))_W = P.$$

Therefore, by definition of univ, we have:

Proposition 3.11. If P computes $f : \mathbb{N} \rightharpoonup \mathbb{N}$ then $\phi_{\widehat{P}} = f$.

Proof. Assume that program P computes f. We want to show that $f = \text{univ}_{\widehat{P}}$, that is

$$f(x) = \text{univ}(\widehat{P}, x)$$

for every $x \in \mathbb{N}$. Indeed:

(1) Assume that $x \in \text{dom} f$. Observe that the execution of $(\widehat{P})_W$ on x is the execution of P on x. The latter returns $f(x)$. On the other hand, the value $\text{univ}(\widehat{P}, x)$ is the value returned by the execution of $(\widehat{P})_W$ on x. Hence $\text{univ}(\widehat{P}, x) = f(x)$.

(2) Assume that $x \notin \text{dom} f$. Then the execution of $(\widehat{P})_W$ on x either does not terminate or it terminates with a result not in $\mathbb{N}$. So $\text{univ}(\widehat{P}, x)$ is undefined. □

Hence, a computable unary function has as many indices as the number of programs that compute f. Thus, the set of indices of a computable unary function is denumerable.

Proposition 3.12. Function univ is a proper universal function.

Proof. Thanks to Proposition 3.2 we know that univ is a computable universal function. It remains to show that univ enjoys the *s-m-n* property. Given $v \in \mathscr{C}_2$, let P_v be a program for computing v. Furthermore, for each $p \in \mathbb{N}$, let P_{v_p} be the program

$$\text{function } (x) \text{ (return } P_v(p,x)).$$

Clearly, for each $p \in \mathbb{N}$, P_{v_p} computes $v_p : \mathbb{N} \rightharpoonup \mathbb{N}$. Hence, by Proposition 3.11,

$$\phi_p^v = v_p = \phi_{\widehat{P_{v_p}}}$$

for every $p \in \mathbb{N}$. Therefore, the desideratum will be fulfilled if we can prove that the map

$$s = \lambda\, p . \widehat{P_{v_p}} : \mathbb{N} \to \mathbb{N}$$

is computable. Indeed, the following algorithm:

$$\text{function } (p) \ (\text{return } P_{J_W}(\text{function}(x)(\text{return } P_v(p,x))))$$

computes s where P_{J_W} is an algorithm that computes J_W. $\qquad\qquad\square$

The existence of proper universal functions is the essential ingredient of a computational model. Namely, it is possible to deal with some operations on computable functions via their indices. This is the case of the following result that provides a binary computable map that mimics, at the level of indices, the composition of computable functions.

Proposition 3.13. Let u be a proper universal function. Then, there exists $c^{\circ} \in m\mathscr{C}_2$ such that

$$\phi_{c^{\circ}(p_1,p_2)}^u = \phi_{p_2}^u \circ \phi_{p_1}^u$$

for every $p_1, p_2 \in \mathbb{N}$.

Proof. Recall maps J, K and L introduced in Notation 1.3. Choose

$$v = \lambda\, p,x . u(L(p), u(K(p),x)) : \mathbb{N}^2 \rightharpoonup \mathbb{N}$$

in order to guarantee

$$(\dagger) \quad (\phi_{p_2}^u \circ \phi_{p_1}^u)(x) = u(p_2, u(p_1,x)) = v(J(p_1,p_2),x)$$

for every $p_1, p_2, x \in \mathbb{N}$. Observe that

$$v = u \circ \langle L \circ \mathrm{proj}_{[1]}^{\mathbb{N}^2}, u \circ \langle K \circ \mathrm{proj}_{[1]}^{\mathbb{N}^2}, \mathrm{proj}_{[2]}^{\mathbb{N}^2} \rangle \rangle.$$

Then, v is computable by Exercise 1.2, Proposition 1.2, Proposition 1.4 and Proposition 1.5. Since u enjoys the *s-m-n* property and v is a computable binary function, let $s \in m\mathscr{C}_1$ be such that

$$v(p,x) = u(s(p),x)$$

for every $p,x \in \mathbb{N}$. Hence, in particular,

$$(\ddagger) \quad v(J(p_1,p_2),x) = u(s(J(p_1,p_2)),x) = \phi_{s(J(p_1,p_2))}^u(x)$$

for every $p_1, p_2, x \in \mathbb{N}$. Thus, from ($\dagger$) and ($\ddagger$),

$$(\phi_{p_2}^u \circ \phi_{p_1}^u)(x) = \phi_{s(J(p_1,p_2))}^u(x)$$

for every $p_1, p_2, x \in \mathbb{N}$, and, so,

$$(\phi_{p_2}^u \circ \phi_{p_1}^u) = \phi_{s(J(p_1,p_2))}^u$$

for every $p_1, p_2 \in \mathbb{N}$. Therefore, the computable map

$$c^\circ = \lambda\, p_1, p_2 . s(J(p_1,p_2))$$

is as required. □

So given indices of two computable functions there is an algorithmic way, given by c°, for obtaining an index for their composition.

Example 3.1. Herein we provide a way of obtaining an index for composition when the proper universal function is univ. Observe that, for each $p_1, p_2 \in \mathbb{N}$, the function

$$\phi_{p_2} \circ \phi_{p_1} : \mathbb{N} \rightharpoonup \mathbb{N}$$

is computed by the program

$$\text{function } (x) \ (\text{return } \mathsf{P}_{\text{univ}}(p_2, \mathsf{P}_{\text{univ}}(p_1, x)))$$

Denote this program by $\mathsf{P}_{p_1 p_2}$. Then, the map

$$c^\circ = \lambda\, p_1, p_2 . \widehat{\mathsf{P}_{p_1 p_2}}$$

is computed by

$$\text{function } (p_1, p_2) \ (\text{return } \mathsf{P}_{J_W}(\mathsf{P}_{p_1 p_2}))$$

and, furthermore, is such that

$$\phi_{c^\circ(p_1,p_2)} = \phi_{\widehat{\mathsf{P}_{p_1 p_2}}} = \phi_{p_2} \circ \phi_{p_1}$$

for every $p_1, p_2 \in \mathbb{N}$ thanks to Proposition 3.11.

The reader may wonder if all operations on computable functions can be dealt with in this way. Clearly, one should expect that only "computable operations" (like composition) can be lifted to the level of indices as computable maps.[2] This issue is addressed with full generality in Section 6.3.

Nevertheless, indices can be used immediately to make precise the notion of computable sequence of computable functions.

---

[2]Note that we use "computable operation" in an intuitive way since our definition of computable does not apply to operations on functions.

Definition 3.9. Let $\lambda k . f_k$ be a sequence of functions in $\mathscr{C}_1$ and u a computable universal function. The sequence is said to be *computable* for u if there is $c \in \mathrm{m}\mathscr{C}_1$ such that, for each $k \in \mathbb{N}$,

$$\phi^u_{c(k)} = f_k.$$

In this context, as usual, we may say computable for computable for univ.

Proposition 3.14. Let u be a universal function and $\lambda k . f_k$ a sequence of computable unary functions. Then

1. If u is computable and $\lambda k . f_k$ is computable for u then $\lambda k,x . f_k(x)$ is computable.

2. If u is proper and $\lambda k,x . f_k(x)$ is computable then $\lambda k . f_k$ is computable for u.

Proof. (1) Assume that u is computable and that $\lambda k . f_k$ is computable for u. Then, there exists a computable unary map c such that, for every $k \in \mathbb{N}$, $c(k)$ is a u-index of f_k, that is, $\phi^u_{c(k)} = f_k$. Hence,

$$\begin{aligned}
\lambda k,x . f_k(x) &= \lambda k,x . \phi^u_{c(k)}(x) \\
&= \lambda k,x . u(c(k),x) \\
&= u \circ \langle c \circ \mathrm{proj}^{\mathbb{N}^2}_{[1]}, \mathrm{proj}^{\mathbb{N}^2}_{[2]} \rangle.
\end{aligned}$$

Hence, $\lambda k,x . f_k(x)$ is a computable function.

(2) Assume that u is proper and that $v = \lambda k,x . f_k(x)$ is computable. As u is a proper universal function, there is a computable map $c : \mathbb{N} \to \mathbb{N}$ such that

$$v(k,x) = u(c(k),x)$$

for every $k,x \in \mathbb{N}$. Hence,

$$f_k(x) = v(k,x) = u(c(k),x) = \phi^u_{c(k)}(x)$$

for every $k,x \in \mathbb{N}$. Thus, $f_k = \phi^u_{c(k)}$, that is, $c(k)$ is a u-index of f_k. As c is computable, we conclude that $\lambda k . f_k$ is computable for u, by definition. $\qquad \Box$

In Section 3.2, capitalizing on the existence of a computable function non-extendable to a computable map (Proposition 3.5), a first example was given of a listable non-decidable set (Proposition 3.6). These two results are easily improved as follows and will be used in Section 4.2.

Proposition 3.15. Let u be a computable universal function. Then,

$$\mathrm{cdiag}^u = \lambda k . \begin{cases} 0 & \text{if } k \in \mathrm{dom}\,\mathrm{diag}^u \text{ and } \mathrm{diag}^u(k) > 0 \\ 1 & \text{if } k \in \mathrm{dom}\,\mathrm{diag}^u \text{ and } \mathrm{diag}^u(k) = 0 \\ \text{undefined} & \text{otherwise.} \end{cases}$$

is a computable unary function with range $\{0,1\}$ which is not extendable to a computable unary map.

Proof. Clearly, cdiag^u is computable and its range is $\{0,1\}$. Assume, by contradiction, that f is a computable map extending cdiag^u. Let $k \in \mathbb{N}$. Consider two cases:
(1) $k \notin \mathrm{dom\,diag}^u$. Since $f(k) \in \mathbb{N}$ then immediately $f(k) \neq \mathrm{diag}^u(k)$.
(2) $k \in \mathrm{dom\,diag}^u$. Then, $f(k) = \mathrm{cdiag}^u(k) \neq \mathrm{diag}^u(k)$.
Hence,
$$f(k) \neq \mathrm{diag}^u(k)$$
for every $k \in \mathbb{N}$, contradicting Proposition 3.4. □

Definition 3.10. Let C, D and E be sets of the same arity. We say that C *separates* D and E if C contains one of them and does not intersect the other.

Clearly, $D \cap E = \emptyset$ whenever C separates D and E. Observe that if C separates D and E then so does C^c.

Exercise 3.4. Show that if two disjoint sets cannot be separated by a decidable set, then none of them is decidable.

Proposition 3.16. Let u be a computable universal function. Then, the sets
$$P^u = \{k : \mathrm{cdiag}^u(k) = 1\} \text{ and } Q^u = \{k : \mathrm{cdiag}^u(k) = 0\}$$
are listable and disjoint but not separated by any decidable set.

Proof. We leave as an exercise the verification of the listability of the two sets. Assume, by contradiction, that a decidable set C separates P^u and Q^u. We have two cases to consider:
(1) $P^u \subseteq C$ and $Q^u \cap C = \emptyset$:
In this case, the computable map χ_C would extend cdiag^u in contradiction with Proposition 3.15.
(2) $Q^u \subseteq C$ and $P^u \cap C = \emptyset$:
In this case, $P^u \subseteq C^c$ and $Q^u \cap C^c = \emptyset$. Thus, the computable map χ_{C^c} would extend cdiag^u in contradiction with Proposition 3.15. □

Observe that the two sets in Proposition 3.16 are both listable but non-decidable (see Exercise 3.4).

3.4 Proper Universal Sets

The concepts and results above on computable functions are now adapted to listable sets, namely to the notion of proper universal set.

Definition 3.11. Given a set $V \in \mathscr{S}_2$, for each $p \in \mathbb{N}$, the unary set

$$V_p = \{x \in \mathbb{N} : (p,x) \in V\}$$

is known as the *section* of V at p.

Exercise 3.5. Show that each section of a listable set is listable.

Definition 3.12. A *proper universal set* is a set $U \in \mathscr{L}_2$ such that for every $V \in \mathscr{L}_2$ there is $s \in \mathrm{m}\mathscr{C}_1$ such that

$$V_p = U_{s(p)}$$

for every $p \in \mathbb{N}$.

That is, $(p,x) \in V$ if and only if $(s(p),x) \in U$ for every $p,x \in \mathbb{N}$. The following result establishes a relationship between proper universal functions and sets.

Proposition 3.17. Let u be a proper universal function. Then, $\mathrm{dom}\,u$ is a proper universal set.

Proof. First, observe that, since u is, by hypothesis, a computable function, $\mathrm{dom}\,u$ is listable, see Proposition 2.11. Let $V \subseteq \mathbb{N}^2$ be an arbitrary listable set. Taking into account the Listability Criteria (Proposition 2.11), with $W = \mathbb{N}$ and $V_2 = \mathbb{N}$, let $g : \mathbb{N}^2 \rightharpoonup \mathbb{N}$ be a computable function with domain V. Since u is, by hypothesis, a proper universal function, there is $s \in \mathrm{m}\mathscr{C}_1$ such that, for every $p \in \mathbb{N}$,

$$g_p = u_{s(p)}$$

and, so, $\mathrm{dom}\,g_p = \mathrm{dom}\,u_{s(p)}$. Hence, for every $p,x \in \mathbb{N}$,

$$(p,x) \in V \text{ iff } (p,x) \in \mathrm{dom}\,g \text{ iff } x \in \mathrm{dom}\,g_p \text{ iff } x \in \mathrm{dom}\,u_{s(p)} \text{ iff } (s(p),x) \in \mathrm{dom}\,u$$

and, so, $\mathrm{dom}\,u$ is a proper universal set since the map s is as required.   □

Remark 3.5. We denote by

$$\mathrm{Univ}$$

the proper universal set $\mathrm{dom}\,\mathrm{univ}$.

We now show that any proper universal set provides an enumeration of the class of all unary listable sets. First we introduce some useful notation.

Remark 3.6. Given a proper universal set U, we denote by

$$\Gamma^U : \mathbb{N} \to \mathscr{L}_1$$

the function $\lambda\,p.U_p$. Moreover, given $p \in \mathbb{N}$, we denote by

$$\Gamma^U_p$$

the set $\Gamma^U(p) = U_p$.

Observe that $\{\Gamma_p^U : p \in \mathbb{N}\} = \{U_p : p \in \mathbb{N}\}$.

Proposition 3.18. Let U be a proper universal set. Then Γ^U is an enumeration of $\mathscr{L}_1$.

Proof. (1) $\{\Gamma_p^U : p \in \mathbb{N}\} \subseteq \mathscr{L}_1$. See Exercise 3.5.

(2) $\mathscr{L}_1 \subseteq \{\Gamma_p^U : p \in \mathbb{N}\}$. Let $A \in \mathscr{L}_1$. We have to find $a \in \mathbb{N}$ such that

$$\Gamma_a^U = A.$$

Consider

$$V = \{0\} \times A.$$

Clearly, $V \in \mathscr{L}_2$ by Proposition 2.4. Furthermore, V_0 coincides with A and every other section of V is empty. Since U is assumed to be proper universal, there is $s \in m\mathscr{C}_1$ such that

$$V_p = U_{s(p)} = \Gamma_{s(p)}^U$$

for each $p \in \mathbb{N}$. Hence, $a = s(0)$ fulfills the requirement, since $A = V_0 = \Gamma_{s(0)}^U$.   □

Definition 3.13. Given a proper universal set U, we say that p is an *U-index* for C whenever $U_p = C$.

Remark 3.7. We may write Γ for Γ^{Univ} and refer to an Univ-index simply as an *index*.

3.5   Solved Problems and Exercises

Problem 3.1. Show that $D = \{p \in \mathbb{N} : p \in \text{range}\,\phi_p\}$ is a listable set.

Solution Observe that for all $q \in \mathbb{N}$,

$$\chi_D^{\text{P}}(q) = 1 \text{ iff } q \in D \text{ iff } q \in \text{range}\,\phi_q \text{ iff } \chi_{\text{range}\,\phi_q}^{\text{P}}(q) = 1 \text{ iff } (\lambda p, x. \chi_{\text{range}\,\phi_p}^{\text{P}}(x))(q,q) = 1.$$

Hence,

$$\chi_D^{\text{P}}(q) = (\lambda p, x. \chi_{\text{range}\,\phi_p}^{\text{P}}(x))(q,q)$$

and so

$$\chi_D^{\text{P}} = (\lambda p, x. \chi_{\text{range}\,\phi_p}^{\text{P}}(x)) \circ \langle \lambda q.q, \lambda q.q \rangle.$$

Then, since $\lambda p, x. \chi_{\text{range}\,\phi_p}^{\text{P}}(x)$ is a computable function (see Exercise 3.6) we can conclude that χ_D^{P} is also a computable function, and so, by Exercise 2.2, D is a listable set.   ▽

Problem 3.2. Let u be a proper universal function. Show that there is $s \in m\mathscr{C}_1$ such that

$$\phi_{s(p)}^u = \lambda x. \phi_p^u(x) + 1,$$

that is, if p is an index of f then $s(p)$ is an index of $\lambda x. f(x) + 1$.

Solution Let $v = \lambda p, x . \phi_p^u(x) + 1$. Observe that

$$\lambda p, x . \phi_p^u(x) + 1 = \lambda p, x . u(p, x) + 1 = (\lambda y . y + 1) \circ u$$

and so v is a computable function since it is the composition of two computable functions. Therefore, $v \in \mathscr{C}_2$. Since u a proper universal function it satisfies the s-m-n property and so there is a map $s \in m\mathscr{C}_1$ such that

$$v_p = \phi_{s(p)}^u$$

for every $p \in \mathbb{N}$. Hence, $\lambda x . \phi_p^u(x) + 1 = \phi_{s(p)}^u$. $\qquad\qquad\qquad \nabla$

Problem 3.3. Let u be a universal computable function and $c \in m\mathscr{C}_2$ such that:

$$\phi_{c(p_1, p_2)}^u = \phi_{p_2}^u \circ \phi_{p_1}^u$$

for every $p_1, p_2 \in \mathbb{N}$. Show that:

1. there is $r \in m\mathscr{C}_1$ such that $\phi_{r(p)}^u = J_p$, for every $p \in \mathbb{N}$;

2. u is a proper function.

Solution (1) Recall that J is the map

$$J = \lambda i, j . \left(\sum_{k=1}^{i+j} k \right) + i = \lambda i, j . i + \frac{1}{2}((i+j)(i+j+1))$$

It is easy to see that $J_{p+1}(k) = J_p(k+1) + 1$ for every $k \in \mathbb{N}$. Thus, denoting by S the successor map $\lambda x . x + 1$, we have:

$$J_{p+1} = S \circ J_p \circ S$$

Since u is a universal function and J_0 is a unary computable map (Exercise 3.1) since J is a computable map (Exercise 1.2), let $p_0 \in \mathbb{N}$ be such that

$$u_{p_0} = J_0.$$

Moreover, because S is a unary computable map let p_S be an u-index of S, that is:

$$u_{p_S} = S.$$

Let $r : \mathbb{N} \to \mathbb{N}$ be the map defined as follows:

$$r(p) = \begin{cases} p_0 & \text{if } p = 0 \\ c(p_S, c(r(p-1), p_S)) & \text{otherwise} \end{cases}.$$

Then:

(a) r is a computable map. Indeed, it is immediate to that r is computed by the following program P_r:

function (p) (if $p == 0$ then return p_0 else return $P_c(p_S, P_c(P_r(p-1), p_S)))$

where P_c is a program that computes c.

(b) $u_{r(p)} = J_p$ for any $p \in \mathbb{N}$ that we prove by induction on p.
(Base) Assume that $p = 0$. Then, $u_{r(0)} = u_{p_0} = J_0$.
(Step) Assume that $p > 0$. Then, by the induction hypothesis, $u_{r(p)} = J_p$. Moreover,

$$u_{r(p+1)} = u_{c(p_S, c(r(p), p_S))} = u_{p_S} \circ u_{c(r(p), p_S)} = S \circ (u_{r(p)} \circ u_{p_S}) = S \circ J_p \circ S = J_{p+1}.$$

(2) u has the s-m-n property. Let $v \in \mathscr{C}_2$ be a function. Observe that

$$v = (v \circ J^{-1}) \circ J$$

and so $v_p = (v \circ J^{-1}) \circ J_p$, for every $p \in \mathbb{N}$. Since u is a universal function and $v \circ J^{-1}$ a unary computable map let $q \in \mathbb{N}$ be such that

$$u_q = v \circ J^{-1}.$$

Hence, $v_p = u_q \circ J_p$ for every $p \in \mathbb{N}$. Using (1), let $r \in \mathfrak{m}\mathscr{C}_1$ be such that $u_{r(p)} = J_p$ for any $p \in \mathbb{N}$. Thus,

$$v_p = u_q \circ u_{r(p)},$$

that is,

$$v_p = \phi^u_{c(r(p), q)}$$

Take $s = c \circ \langle r, (\lambda x . q) \rangle : \mathbb{N} \to \mathbb{N}$. It follows straightforwardly that s is a computable map. $\triangledown$

Problem 3.4. Let u be a proper universal function and $v \in \mathscr{C}_2$. Prove that $\lambda p . v_p$ is computable for u. Conclude that, in particular, $\lambda k . \phi^u_k$ is computable for u. Verify that the latter assertion still holds if u is only assumed to be computable and universal.

Solution (1) To show that $\lambda p . v_p$ is computable for u we need to find a map $c \in \mathfrak{m}\mathscr{C}_1$ such that $v_p = \phi^u_{c(p)}$ for each $p \in \mathbb{N}$. The existence of such a map is an immediate consequence of u being a proper universal function.

(2) The fact that $\lambda k . \phi^u_k$ is also computable for u is an immediate consequence of $u \in \mathscr{C}_2$ and (1).

(3) Assume that u is universal and computable. Take $c \in \mathfrak{m}\mathscr{C}_1$ to be the identity map. Then, it is clear that $\lambda k . \phi^u_k$ is computable for u. $\triangledown$

Problem 3.5. Provide an example of a non-computable function in $\mathscr{F}_1$ with listable but non-decidable domain.

Solution We start by defining a bijection between $\mathbb{K}$ and $\mathbb{K}^c$. Consider the following injective enumeration $h_{\mathbb{K}^c} : \mathbb{N} \to \mathbb{N}$ of $\mathbb{K}^c$ (necessarily non-computable since $\mathbb{K}^c$ is non-listable):

- $h_{\mathbb{K}^c}(0) = \mu i . i \notin \mathbb{K}$;

- $h_{\mathbb{K}^c}(k+1) = \mu i . i \notin (\mathbb{K} \cup \{h_{\mathbb{K}^c}(0), \ldots, h_{\mathbb{K}^c}(k)\})$.

Let $h_{\mathbb{K}}$ be a computable injective enumeration of $\mathbb{K} = \mathbb{K}^{\text{univ}}$ (that exists by Proposition 2.9 given that $\mathbb{K}$ is listable). Consider the bijection $b : \mathbb{K} \to \mathbb{K}^c$ such that

$$b = h_{\mathbb{K}^c} \circ h_{\mathbb{K}}^{-1}.$$

It is worth mentioning that any such bijection between $\mathbb{K}$ and $\mathbb{K}^c$ is necessarily non-computable. Otherwise it would be possible to define a computable enumeration of $\mathbb{K}^c$, a set that we know to be non-listable (Proposition 3.7).

Consider the function

$$f = \lambda x . \begin{cases} b(x) & \text{if } x \in \mathbb{K} \\ \text{undefined} & \text{otherwise} \end{cases} : \mathbb{N} \rightharpoonup \mathbb{N}.$$

Observe that $\text{dom} f = \mathbb{K}$. Furthermore f is not computable. Indeed, if f was a computable function then

$$f \circ h_{\mathbb{K}} = h_{\mathbb{K}^c} \circ h_{\mathbb{K}}^{-1} \circ h_{\mathbb{K}} = h_{\mathbb{K}^c}$$

would also be a computable function. Therefore, f is an example of a non-computable function in $\mathscr{F}_1$ with listable non-decidable domain $\mathbb{K}$.   ▽

Problem 3.6. Let u be a proper universal function and U a proper universal set. Show that there is a map $s \in \text{m}\mathscr{C}_1$ such that $u_{s(p)} = \chi_{U_p}^{\text{P}}$, for every $p \in \mathbb{N}$.

Solution Observe that $\chi_U^{\text{P}} \in \mathscr{C}_2$, since $U \subseteq \mathbb{N}^2$ is a listable set. Hence, because u is a proper universal function, there is $s \in \text{m}\mathscr{C}_1$ such that

$$(\chi_U^{\text{P}})_p = u_{s(p)},$$

for every $p \in \mathbb{N}$, by the s-m-n property of u. Hence, for every $p \in \mathbb{N}$,

$$(\chi_U^{\text{P}})_p(x) = 1 \text{ iff } (\chi_U^{\text{P}})(p,x) = 1 \text{ iff } (p,x) \in U \text{ iff } x \in U_p \text{ iff } \chi_{U_p}^{\text{P}}(x) = 1.$$

Thus, $(\chi_U^{\text{P}})_p = \chi_{U_p}^{\text{P}}$ and so $u_{s(p)} = \chi_{U_p}^{\text{P}}$, for every $p \in \mathbb{N}$.   ▽

Problem 3.7. Let u' be a proper universal function, $U \subseteq \mathbb{N}^2$ a listable set and $b \in \mathfrak{m}\mathscr{C}_1$ a bijection such that $U_p = (\text{dom}\, u')_{b(p)}$ for each $p \in \mathbb{N}$. Then, there is a proper universal function u such that $\text{dom}\, u = U$.

Solution Let u be the function

$$\lambda\, p, x . u'(b(p), x).$$

Clearly, if b computable and is a bijection then b^{-1} is a bijection and is computable.

(1) We start by showing that u is a proper universal function. Let $v \in \mathscr{C}_2$. Observe that there is a computable map $s' : \mathbb{N} \to \mathbb{N}$ such that $v(p,x) = u'(s'(p), x)$, since u' is a proper universal function for every $p, x \in \mathbb{N}$. But, by definition, $u'(p,x) = u(b^{-1}(p), x)$ for every $p, x \in \mathbb{N}$. Hence,

$$v(p,x) = u(b^{-1}(s'(p)), x)$$

for every $p, x \in \mathbb{N}$. Take $s = b^{-1} \circ s'$. Then s is a computable map and $v_p = u_{s(p)}$ for every $p \in \mathbb{N}$. So we conclude that u enjoys the *s-m-n* property.

(2) It remains to check that $\text{dom}\, u = U$. In fact,

$$
\begin{array}{llll}
(p,x) \in \text{dom}\, u & \text{iff} & (b(p), x) \in \text{dom}\, u' \\
& \text{iff} & x \in (\text{dom}\, u')_{b(p)} \\
& \text{iff} & x \in U_p \\
& \text{iff} & (p,x) \in U
\end{array}
$$

for every $p, x \in \mathbb{N}$. $\qquad\qquad \nabla$

Problem 3.8. Show that there is $r \in \mathfrak{m}\mathscr{C}_2$ such that $\text{Univ}_{r(p,q)} = \phi_p^{-1}(\text{Univ}_q)$ for every $p, q \in \mathbb{N}$.

Solution Let

$$V = \{(p,q,x) \in \mathbb{N}^3 : x \in \phi_p^{-1}(\text{Univ}_q)\}.$$

Observe that

$$
\begin{array}{rl}
V = & \{(p,q,x) \in \mathbb{N}^3 : x \in \phi_p^{-1}(\text{Univ}_q)\} \\
= & \{(p,q,x) \in \mathbb{N}^3 : \phi_p(x) \in \text{Univ}_q\} \\
= & \{(p,q,x) \in \mathbb{N}^3 : \text{univ}_p(x) \in \text{Univ}_q\} \\
= & \{(p,q,x) \in \mathbb{N}^3 : \chi^p_{\text{Univ}_q}(\text{univ}_p(x)) = 1\} \\
= & \{(p,q,x) \in \mathbb{N}^3 : \chi^p_{\text{Univ}}(q, \text{univ}_p(x)) = 1\} \\
= & \{(p,q,x) \in \mathbb{N}^3 : \chi^p_{\text{Univ}}(q, \text{univ}(p,x)) = 1\}.
\end{array}
$$

Thus,

$$\chi^p_V = \chi^p_{\text{Univ}} \circ \langle \text{proj}^{\mathbb{N}^3}_{[2]}, \text{univ} \circ \langle \text{proj}^{\mathbb{N}^3}_{[1]}, \text{proj}^{\mathbb{N}^3}_{[3]} \rangle \rangle.$$

Hence, V is listable because its characteristic function is computable. Consider now the set

$$V' = \{(J(p,q),x) : (p,q,x) \in V\}.$$

Then,

$$\chi_{V'}^{\mathrm{p}} = \chi_V^{\mathrm{p}} \circ \langle K \circ \mathrm{proj}_{[1]}^{\mathbb{N}^2}, L \circ \mathrm{proj}_{[1]}^{\mathbb{N}^2}, \mathrm{proj}_{[2]}^{\mathbb{N}^2} \rangle.$$

Therefore, V' is listable because its characteristic function is computable. Observe that

$$x \in \phi_p^{-1}(\mathrm{Univ}_q) \quad \text{iff} \quad (p,q,x) \in V \quad \text{iff} \quad (J(p,q),x) \in V',$$

for every $p,q,x \in \mathbb{N}$. Moreover, taking into account that Univ is a proper universal set, there is $s \in \mathrm{m}\mathscr{C}_1$ such that,

$$(J(p,q),x) \in V' \quad \text{iff} \quad (s(J(p,q)),x) \in \mathrm{Univ},$$

for every $p,q,x \in \mathbb{N}$. Therefore, the computable binary map

$$r = \lambda\, p,q \cdot s(J(p,q))$$

is such that

$$x \in \phi_p^{-1}(\mathrm{Univ}_q) \text{ iff } (J(p,q),x) \in V' \text{ iff } (r(p,q),x) \in \mathrm{Univ} \text{ iff } x \in \mathrm{Univ}_{r(p,q)}$$

for every $p,q,x \in \mathbb{N}$. Hence, $\phi_p^{-1}(\mathrm{Univ}_q) = \mathrm{Univ}_{r(p,q)}$ for every $p,q,x \in \mathbb{N}$. ∇

Problem 3.9. Show that there is no function universal for $\mathscr{F}_1$ using cardinality arguments.

Solution Observe that

$$|\mathbb{N}| \geq |S_u|$$

where S_u is the set of all sections of a binary function u. On the other hand,

$$|\mathscr{F}_1| > |\mathbb{N}|.$$

Thus, $|\mathscr{F}_1| > |\mathbb{N}| \geq |S_u|$ and so there are not enough sections of u to represent all the functions in $\mathscr{F}_1$.

∇

Supplementary Exercises

Exercise 3.6. Show that $\lambda\, p,x \cdot \chi_{\mathrm{range}\,\phi_p}^{\mathrm{p}}(x)$ is a computable function.

Exercise 3.7. Show that $\mathrm{univ} \circ \langle S \circ S \circ \mathrm{proj}_{[1]}^{\mathbb{N}^2}, \mathrm{proj}_{[2]}^{\mathbb{N}^2} \rangle$ is a proper universal function.

Exercise 3.8. Let u be a computable universal function and $c \in \mathbb{N}$. Show that $\{p \in \mathbb{N} : \phi_p^u(p) = c\}$ is a listable set.

Exercise 3.9. Let u be a proper universal computable function. Show that there is $c \in m\mathscr{C}_2$ such that

$$\phi_{c(p_1,p_2)}^u = \lambda x . \phi_{p_1}^u(x) + \phi_{p_2}^u(x).$$

That is, if p_1 and p_2 are indices of f_1 and f_2, respectively, then $c(p_1, p_2)$ is the index of $f_1 + f_2$.

Exercise 3.10. Let

$$\text{univ}_W : \mathbb{N} \times W \rightharpoonup W$$

be the function computed by the following program P_{univ_W}:

$$\text{function } (p, w) \ (\text{return } P_W(p)(w))$$

where P_W is a program that computes $\lambda k . k_W$ (recall Exercise 1.8). So

$$\text{univ}_W = \lambda p, w . \begin{cases} w' & \text{if } w' \text{ results from the execution of } p_W \text{ on } w \\ \text{undefined} & \text{otherwise} \end{cases} .$$

Moreover, denote by $\mathscr{C}_1^W$ the set of computable functions from W to W. Show that function univ_W is universal with respect to $\mathscr{C}_1^W$.

Exercise 3.11. Let U be a proper universal set and u a proper universal function. Show that there exists $c \in m\mathscr{C}_1$ such that

$$U_{c(p)} = \text{dom } \phi_p^u$$

for every $p \in \mathbb{N}$.

Exercise 3.12. Show that there is $s \in m\mathscr{C}_1$ such that $\Gamma_{s(p)} = J(\text{graph } \phi_p)$ for every $p \in \mathbb{N}$.

Exercise 3.13. Let $u \in \mathscr{F}_2$ be a function. Show that, for every $p \in \mathbb{N}$,

$$p \in \mathbb{K}^u \qquad \text{iff} \qquad p \in \Gamma_p^{\text{dom } u}.$$

Exercise 3.14. Let U be a proper universal set and u a proper universal function. Prove or refute the existence of the following maps:

1. $c^\cap \in m\mathscr{C}_2$ such that, for every $p_1, p_2 \in \mathbb{N}$,

$$\Gamma_{c^\cap(p_1,p_2)}^U = \Gamma_{p_1}^U \cap \Gamma_{p_2}^U;$$

2. $c^\times \in \mathrm{m}\mathscr{C}_2$ such that, for every $p_1, p_2 \in \mathbb{N}$,

$$\Gamma^U_{c^\times(p_1,p_2)} = J(\Gamma^U_{p_1} \times \Gamma^U_{p_2});$$

3. $c^c \in \mathrm{m}\mathscr{C}_1$ such that, for every $p \in \mathbb{N}$,

$$\Gamma^U_{c^c(p)} = (\Gamma^U_p)^c;$$

4. $c^f \in \mathrm{m}\mathscr{C}_1$, for any given $f \in \mathscr{C}_1$, such that, for every $p \in \mathbb{N}$,

$$\Gamma^U_{c^f(p)} = f(\Gamma^U_p);$$

5. $c^{f^{-1}} \in \mathrm{m}\mathscr{C}_1$, for any given $f \in \mathscr{C}_1$, such that, for every $p \in \mathbb{N}$,

$$\Gamma^U_{c^{f^{-1}}(p)} = f^{-1}(\Gamma^U_p);$$

6. $c^{\mathrm{dom}} \in \mathrm{m}\mathscr{C}_1$ such that, for every $p \in \mathbb{N}$,

$$\Gamma^U_{c^{\mathrm{dom}}(p)} = \mathrm{dom}\, \phi^u_p.$$

Exercise 3.15. Let u be a universal function and t the function $\lambda\, p,x,y\,.\,u(p,J(x,y))$.

1. Show that t is universal for $\mathscr{C}_2$;

2. Show that $u' = \lambda\, p,x\,.\,t(K(p),L(p),x)$ is a proper universal function.

Exercise 3.16. Let $\lambda k.\, f_k$ be a computable sequence of functions in $\mathscr{C}_1$. Show that

$$\left(\bigcup_{k\in\mathbb{N}} \mathrm{dom}\, f_k \right) \in \mathscr{L}_1$$

Exercise 3.17. Let u be a computable universal function and $\lambda\, k.\, f_k$ a sequence of computable unary functions computable for u. Prove that

$$\bigcup_{k\in\mathbb{N}} \mathrm{range}\, f_k \in \mathscr{L}_1.$$

Provide a counterexample in order to show that the assumption of the computability for u of the sequence is essential.

Chapter 4

Theorems on Function Indices

In this chapter we analyze decidability and listability of sets of indices of computable unary functions.

4.1 Rice's Theorem

Let us return to the issue raised in Section 3.3: for a given universal function u, what is the nature of the set of u-indices of a given computable unary function f? The following results answer this question for the particular case of the undefined unary function.

Proposition 4.1. Let u be a proper universal function. Then:

$$\{p \in \mathbb{N} : \operatorname{dom} \phi_p^u = \emptyset\} \notin \mathcal{D}.$$

Proof. The proof is carried out using the problem reduction technique described in Section 2.3. Observe that the problem

$$\text{Given } q \in \mathbb{N}, \text{ does } q \in \mathbb{K}^c?$$

can be reduced to the problem

$$\text{Given } q \in \mathbb{N}, \text{ does } q \in \{p \in \mathbb{N} : \operatorname{dom} \phi_p^u = \emptyset\}?$$

Let

$$v = \lambda \, q, x . \begin{cases} 1 & \text{if } q \in \mathbb{K} \\ \text{undefined} & \text{otherwise} \end{cases} : \mathbb{N}^2 \rightharpoonup \mathbb{N}.$$

Note that

$$v = \lambda \, q, x . \chi_{\mathbb{K}}^{\mathrm{P}}(q)$$

79

that is

$$v = \chi_{\mathbb{K}}^{\mathrm{p}} \circ \mathrm{proj}_{[1]}^{\mathbb{N}^2}.$$

Hence, v is computable since $\mathbb{K}$ is listable by Proposition 3.6. Moreover, the sections of v are such that:

- v_q is $\lambda x. 1$ whenever $q \in \mathbb{K}$;

- v_q is $\lambda x.$ undefined whenever $q \notin \mathbb{K}$.

Since u is assumed to be a proper universal function, u enjoys the *s-m-n* property and, so, there is a computable unary map s such that

$$v_q = u_{s(q)} = \phi_{s(q)}^u$$

for every $q \in \mathbb{N}$. Thus,

- $\phi_{s(q)}^u$ is $\lambda x. 1$ whenever $q \in \mathbb{K}$;

- $\phi_{s(q)}^u$ is $\lambda x.$ undefined whenever $q \notin \mathbb{K}$.

We now show that

$$q \in \mathbb{K}^{\mathrm{c}} \quad \text{iff} \quad \mathrm{dom}\, \phi_{s(q)}^u = \emptyset$$

for every $q \in \mathbb{N}$.

($\rightarrow$) Assume that $q \in \mathbb{K}^{\mathrm{c}}$. Then $\phi_{s(q)}^u$ is $\lambda x.$ undefined. Hence, $\mathrm{dom}\, \phi_{s(q)}^u = \emptyset$ and so $s(q) \in \{p \in \mathbb{N} : \mathrm{dom}\, \phi_p^u = \emptyset\}$.

($\leftarrow$) Assume that $q \in \mathbb{K}$. Then $\phi_{s(q)}^u$ is $\lambda x. 1$. Therefore, $\mathrm{dom}\, \phi_{s(q)}^u \neq \emptyset$ and so $s(q) \notin \{p \in \mathbb{N} : \mathrm{dom}\, \phi_p^u = \emptyset\}$.

So, the first problem reduces to the second problem.

Hence, by Proposition 2.14, the set $\{p \in \mathbb{N} : \mathrm{dom}\, \phi_p^u = \emptyset\}$ is non-decidable since $\mathbb{K}^{\mathrm{c}}$ is not decidable, by Proposition 3.2 and Proposition 3.7. $\qquad\square$

Since every finite set is decidable, Proposition 4.1 implies that, for every proper universal function u, the set of u-indices of the undefined function is infinite and, hence, denumerable.

The interested reader will be able to adapt the proof of Proposition 4.1 in order to show that, for every proper universal function, the set of indices of the undefined function, besides being non-decidable, is also non-listable. This non-listability result is obtained below (Proposition 4.3) instead as an immediate corollary of Proposition 4.1 and the following useful positive result.

Proposition 4.2. Let u be a computable universal function. Then:

$$\{p \in \mathbb{N} : \mathrm{dom}\, \phi_p^u \neq \emptyset\} \in \mathscr{L}.$$

Proof. Observe that $\{p \in \mathbb{N} : \operatorname{dom} \phi_p^u \neq \emptyset\}$ is a non-empty set. Let q be one of its elements. It is immediate to see that the following algorithm

```
function (k) (
    p,x,t = Pβ³₁(k);
    if isnat(stceval(Pu,p,x,t,−1)) then
        return p
    else
        return q
)
```

computes an enumeration for $\{p \in \mathbb{N} : \operatorname{dom} \phi_p^u \neq \emptyset\}$ where $\mathsf{P}_{\beta_1^3}$ and P_u are programs that compute β_1^3 and u, respectively. ☐

Proposition 4.3. Let u be a proper universal function. Then:

$$\{p \in \mathbb{N} : \operatorname{dom} \phi_p^u = \emptyset\} \notin \mathscr{L}.$$

Proof. Immediate corollary of Proposition 4.1 and Proposition 4.2, taking into account Post's Theorem (Proposition 2.6). ☐

Proposition 4.1 above is a special case of Rice's Theorem (Proposition 4.4) that provides a necessary and sufficient condition for the decidability of the set of indices of a class of computable functions, assuming that the universal function at hand is proper. To this end, it is convenient to introduce some additional notation.

Remark 4.1. Given $A \subseteq \mathscr{C}_1$ and a universal function u, we denote by

$$\operatorname{ind}^u(A)$$

the set $\{p \in \mathbb{N} : \phi_p^u \in A\}$ of u-indices of the functions in A. We may write

$$\operatorname{ind}(A)$$

for $\operatorname{ind}^{\operatorname{univ}}(A)$. Finally, given $f \in \mathscr{C}_1$, we denote by

$$\operatorname{ind}^u(f)$$

the set $\operatorname{ind}^u(\{f\})$.

Exercise 4.1. Show that $\operatorname{ind}^u(\mathscr{C}_1 \setminus A) = (\operatorname{ind}^u(A))^c$, for each $A \subseteq \mathscr{C}_1$.

Proposition 4.4 (Rice's Theorem). Let $A \subseteq \mathscr{C}_1$ and u be a proper universal function. Then:

$$\operatorname{ind}^u(A) \in \mathscr{D} \quad \text{iff} \quad A = \emptyset \text{ or } A = \mathscr{C}_1.$$

Proof. Clearly, if $A = \emptyset$ or $A = \mathscr{C}_1$ then $\mathrm{ind}^u(A)$ is decidable. It remains to prove that if $A \neq \emptyset$ and $A \neq \mathscr{C}_1$ then $\mathrm{ind}^u(A)$ is non-decidable.

Recall that a set is decidable if and only if its complement is decidable, see Proposition 2.1. So, without loss of generality, we may assume that the undefined function is in A (otherwise we just adapt the proof to the case $B = \mathscr{C}_1 \setminus A$ and prove that $\mathrm{ind}^u(B) = (\mathrm{ind}^u(A))^c$ is non-decidable).

The proof mimics the proof of Proposition 4.1 with minor adaptations. Under the assumption $\emptyset \subsetneq A \subsetneq \mathscr{C}_1$, we show that if $\mathrm{ind}^u(A)$ is decidable then so is $\mathbb{K}^c$, in contradiction with Proposition 3.6.

Let $g : \mathbb{N} \rightharpoonup \mathbb{N}$ be an element of $\mathscr{C}_1 \setminus A$. Consider the binary function

$$v = \lambda\, q,x. \begin{cases} g(x) & \text{if } q \in \mathbb{K} \\ \text{undefined} & \text{otherwise.} \end{cases}$$

Observe that

$$v = (\lambda\, x,y\,.\,x{\times}y) \circ \langle g \circ \mathrm{proj}^{\mathbb{N}^2}_{[2]}, \chi^{\mathrm{p}}_{\mathbb{K}} \circ \mathrm{proj}^{\mathbb{N}^2}_{[1]} \rangle.$$

Hence, v is computable. Since u is a proper universal function then there is $s \in \mathrm{m}\mathscr{C}_1$ such that $v_q = \phi^u_{s(q)}$ for every $q \in \mathbb{N}$. Hence,

- $\phi^u_{s(q)}$ is g whenever $q \in \mathbb{K}$;

- $\phi^u_{s(q)}$ is the undefined function whenever $q \notin \mathbb{K}$.

Hence, since the undefined function is in A and g was chosen in $\mathscr{C}_1 \setminus A$, we have:

$$q \in \mathbb{K}^c \quad \text{iff} \quad s(q) \in \mathrm{ind}^u(A)$$

and, so, the decidability of $\mathrm{ind}^u(A)$ would entail the decidability of $\mathbb{K}^c$ contradicting Proposition 3.6.   □

Rice's theorem is quite useful when we want to establish the undecidability of a set or problem as illustrated in the following exercise.

Exercise 4.2. Let u be a proper universal function and $k \in \mathbb{N}$. Show that the following problems are non-decidable:

1. Given $p \in \mathbb{N}$, is $\phi^u_p = \lambda\, x.k$?

2. Given $p \in \mathbb{N}$, does $k \in \mathrm{dom}\,\phi^u_p$?

3. Given $p \in \mathbb{N}$, does $k \in \mathrm{range}\,\phi^u_p$?

However, it should be stressed that the undecidability of a set or problem concerning indices with respect to a proper universal function is not always obtainable as a direct application of Rice's theorem. Some additional work may be necessary or Rice's Theorem may even be useless, as the reader is asked to discover in the next exercise.

Exercise 4.3. Let u be a proper universal function. Show that the following sets are non-decidable:

1. $\{(p,q) \in \mathbb{N}^2 : \phi_p^u = \phi_q^u\}$;

2. $\{p \in \mathbb{N} : p \in \mathrm{dom}\, \phi_p^u\}$.

Rice's Theorem does help in answering the following question: are all computable universal functions proper ones? The reader is invited to reach the (negative) answer to this question in the following exercise.

Exercise 4.4. Show that there exists a non-proper computable universal function.

4.2   Rice-Shapiro Theorem

Rice's Theorem can be obtained from the more general result below (attributed to Henry Gordon Rice and Stewart Shapiro) that provides a necessary condition for the listability of the set of indices of a class of computable functions, again when we are dealing with a proper universal function. Recall Definition 1.6. We start by stating some properties of the subfunction relation.

Exercise 4.5. Show that $\subseteq$ is a partial order (i.e. a relation that is reflexive, transitive and antisymmetric) in $\mathscr{F}_1$ and in $\mathscr{C}_1$. Moreover, show that $f_1 \subseteq f_2$ if and only if graph $f_1 \subseteq$ graph f_2.

To prove the Rice-Shapiro Theorem (Proposition 4.5) we need some additional terminology and notation.

Definition 4.1. By a *finite function* we understand a function ϑ such that $\mathrm{dom}\, \vartheta$ is a finite set.

Recall that every finite function is computable (see Problem 1.1).

Remark 4.2. Given $n \in \mathbb{N}^+$, we denote by

$$\mathrm{f}\mathscr{C}_n$$

the set of finite n-ary functions.

Proposition 4.5 (Rice-Shapiro Theorem). Let $A \subseteq \mathscr{C}_1$ and u be a proper universal function. Then, the Rice-Shapiro condition

$$\forall f \in \mathscr{C}_1 : (f \in A \quad \text{iff} \quad \exists \text{finite } \vartheta \subseteq f : \vartheta \in A) \qquad \text{(RSC)}$$

holds whenever $\text{ind}^u(A) \in \mathscr{L}$.

Proof. The proof is carried out by counterposition. Assume that the Rice-Shapiro condition fails. There are two cases:

(1) there is $f \in A$ such that $\forall$ finite $\vartheta \subseteq f : \vartheta \notin A$;

(2) there are $f \in \mathscr{C}_1$ and $\vartheta \in f\mathscr{C}_1$ such that $f \notin A$, $\vartheta \in A$ and $\vartheta \subseteq f$.

In both cases, we are able to reduce the problem

$$\text{Given } p \in \mathbb{N}, \text{ does } p \in (\mathbb{K}^u)^{\complement}?$$

to the problem

$$\text{Given } p \in \mathbb{N}, \text{ does } p \in \text{ind}^u(A)?$$

as we proceed to show.

(1) Let $f \in A$ be such that $\forall$ finite $\vartheta \subseteq f : \vartheta \notin A$.

Consider the binary function v computed by the following procedure:

```
function (p,t) (
    if isnat(stceval(P_{χ_{𝕂u}^p}, p, t, −1)) then
        while true do null
    else
        return P_f(t)
)
```

where $P_{\chi_{\mathbb{K}u}^p}$ and P_f are programs that compute $\chi_{\mathbb{K}u}^p$ and f, respectively. Observe that, by construction of v, we have

$$(\S) \qquad v_p \subseteq f$$

for every $p \in \mathbb{N}$. Furthermore, if $p \in \mathbb{K}^u$ there is $t \in \mathbb{N}$ such that the execution of $P_{\chi_{\mathbb{K}u}^p}$ on p terminates in time less or equal to t time units. Hence, if $p \in \mathbb{K}^u$ there is $t \in \mathbb{N}$ such that $(p,t') \notin \text{dom } v$ and, so, $t' \notin \text{dom } v_p$, for every $t' \geq t$. That is,

$$p \in \mathbb{K}^u \quad \text{implies} \quad v_p \in f\mathscr{C}_1$$

and, so, taking into account (1) as well as ($\S$),

$$(\dagger) \qquad p \in \mathbb{K}^u \quad \text{implies} \quad v_p \notin A.$$

On the other hand, again by construction of v,

$$p \notin \mathbb{K}^u \quad \text{implies} \quad v_p = f$$

and, thus, taking into account (1),

$$(\dagger\dagger) \quad p \notin \mathbb{K}^u \quad \text{implies} \quad v_p \in A.$$

Since u is proper universal, there is $s \in \mathrm{m}\mathscr{C}_1$ such that

$$v_p = \phi^u_{s(p)}$$

for every $p \in \mathbb{N}$ and, so:

$$p \in \mathbb{K}^u \quad \text{implies} \quad \phi^u_{s(p)} \notin A, \text{ thanks to } (\dagger);$$
$$p \notin \mathbb{K}^u \quad \text{implies} \quad \phi^u_{s(p)} \in A, \text{ thanks to } (\dagger\dagger).$$

That is,

$$p \in (\mathbb{K}^u)^{\complement} \text{ iff } s(p) \in \mathrm{ind}^u(A)$$

for every $p \in \mathbb{N}$. Therefore, as envisaged, we obtain the requested reduction.

(2) Let $f \in \mathscr{C}_1$ and $\vartheta \in f\mathscr{C}_1$ be such that $f \notin A$, $\vartheta \in A$ and $\vartheta \subseteq f$.
Consider now the binary function v' computed by the following program:

```
function (p,x) (
    if P_{χ_dom ϑ}(x) == 1 then
        return P_f(x)
    else
        if P_{χ^p_{K^u}}(p) == 1 then
            return P_f(x)
        else
            while true do null
)
```

where $P_{\chi^p_{\mathbb{K}^u}}$, $P_{\chi_{\mathrm{dom}\,\vartheta}}$ and P_f are programs that compute $\chi^p_{\mathbb{K}^u}$, $\chi_{\mathrm{dom}\,\vartheta}$ and f, respectively.
Observe that, by construction of v', we have

$$p \in \mathbb{K}^u \quad \text{implies} \quad v'_p = f$$

and, thus, since $f \notin A$ by (2),

$$(\dagger') \quad p \in \mathbb{K}^u \quad \text{implies} \quad v'_p \notin A.$$

On the other hand, again by construction of v', we have

$$p \notin \mathbb{K}^u \quad \text{implies} \quad \forall x \in \mathbb{N} \begin{cases} x \in \mathrm{dom}\,\vartheta & \text{implies} \quad v'_p(x) = v'(p,x) = f(x) \\ x \notin \mathrm{dom}\,\vartheta & \text{implies} \quad v'_p(x) = v'(p,x) = \text{undefined} \end{cases}$$

and, hence, since $\vartheta \subseteq f$ by (2),

$$p \notin \mathbb{K}^u \quad \text{implies} \quad v'_p = \vartheta.$$

So, since $\vartheta \in A$ by (2),

$$(\dagger\dagger') \quad p \notin \mathbb{K}^u \quad \text{implies} \quad v'_p \in A.$$

Recalling that u is assumed to be proper universal, there is $s' \in m\mathscr{C}_1$ such that

$$v'_p = \phi^u_{s'(p)}$$

for every $p \in \mathbb{N}$. So

$$p \in \mathbb{K}^u \quad \text{implies} \quad \phi^u_{s'(p)} \notin A, \text{ thanks to } (\dagger');$$

$$p \notin \mathbb{K}^u \quad \text{implies} \quad \phi^u_{s'(p)} \in A, \text{ thanks to } (\dagger\dagger').$$

That is,

$$p \in (\mathbb{K}^u)^{\complement} \quad \text{iff} \quad s'(p) \in \text{ind}^u(A).$$

Thus, as envisaged, we obtain the requested reduction.

Hence, by Proposition 2.14, if $\text{ind}^u(A)$ were listable then $(\mathbb{K}^u)^{\complement}$ would also be listable contradicting Proposition 3.7. $\qquad\qquad\square$

Observe that the necessary condition (RSC) for the listability of $\text{ind}^u(A)$ provided by Proposition 4.5 is not a sufficient condition. The reader is asked to provide a counterexample in Exercise 4.25, after a necessary and sufficient condition for the listability of $\text{ind}^u(A)$ is established in Proposition 4.8.

Exercise 4.6. Let u be a proper universal function and $\text{ind}^u(A) \in \mathscr{L}_1$. Show that:

1. If $A \neq \emptyset$ then there is a finite function in A.

2. If the undefined function is in A then $A = \mathscr{C}_1$ and so $\text{ind}^u(A)$ is a decidable set.

3. Any extension of a function in A is also in A.

When working with a proper universal function, the Rice-Shapiro Theorem may be quite useful for proving that a set of indices is non-listable, as illustrated in the following example, or that a problem concerning indices is not semidecidable.

Example 4.1. We now illustrate the Rice-Shapiro Theorem by showing that the set

$$\{p \in \mathbb{N} : \phi^u_p \text{ is not total}\}$$

is non-listable. Indeed, consider the set

$$A = \{f \in \mathscr{C}_1 : f \text{ is not total}\}.$$

Then, $\mathrm{ind}^u(A) = \{p \in \mathbb{N} : \phi_p^u \text{ is not total}\}$. Set A is clearly not empty as any finite function is in A. Let $\vartheta \in A$ be a finite function and consider the following extension:

$$f = \lambda x. \begin{cases} \vartheta(x) & \text{if } x \in \mathrm{dom}\, \vartheta \\ 0 & \text{otherwise.} \end{cases}$$

Observe that f is computable because ϑ is computable and $\mathrm{dom}\, \vartheta$ is decidable because it is finite. Indeed, $f = \lambda x.\, \vartheta(x) \times \chi_{\mathrm{dom}\,\vartheta}(x)$. By construction, f is an extension of ϑ, that is, $\vartheta \subseteq f$, and f is total, that is, $f \notin A$. So, A does not fulfill the Rice-Shapiro condition and, consequently, $\mathrm{ind}^u(A)$ is non-listable.

Exercise 4.7. Obtain Rice's Theorem as a corollary of the Rice-Shapiro Theorem.

Observe that for any proper universal function u, Rice's Theorem entails that the set $\mathrm{ind}^u(f)$, is non-decidable and, so, infinite. Moreover, taking into account the Rice-Shapiro Theorem, there is no hope, in general, of a computable enumeration of $\mathrm{ind}^u(f)$. However, the following result shows that there is a computable enumeration of an infinite subset of $\mathrm{ind}^u(f)$.

Proposition 4.6. Let u be a proper universal function. Then, given $f \in \mathscr{C}_1$, there is $r \in \mathrm{m}\mathscr{C}_1$ such that:

- r is injective;

- $\mathrm{range}\, r \subseteq \mathrm{ind}^u(f)$.

Proof. Let $f \in \mathscr{C}_1$. It is convenient to consider two cases:

(1) $f \neq \lambda x.\,\mathrm{undefined}$. Consider the function:

$$g^\dagger = \lambda\, p, x. \begin{cases} f(x) & \text{if } p \in \mathbb{K}^u \\ \mathrm{undefined} & \text{otherwise} \end{cases} : \mathbb{N}^2 \to \mathbb{N}.$$

Since $\mathbb{K}^u$ is listable, function $g^\dagger$ is computable because

$$g^\dagger = \lambda\, p, x.\, \chi_{\mathbb{K}^u}^{\mathrm{P}}(p) \times f(x) = (\lambda\, x_1, x_2.\, x_1 \times x_2) \circ \langle \chi_{\mathbb{K}^u}^{\mathrm{P}} \circ \mathrm{proj}_{[1]}^{\mathbb{N}^2}, f \circ \mathrm{proj}_{[2]}^{\mathbb{N}^2} \rangle.$$

Clearly, for each $p \in \mathbb{N}$,

- $p \in \mathbb{K}^u$ implies $g_p^\dagger = f$;

- $p \notin \mathbb{K}^u$ implies $g_p^\dagger = \lambda x.\,\mathrm{undefined}$.

Furthermore, since u is a proper universal function, there is $s^\dagger \in \mathrm{m}\mathscr{C}_1$ such that:

$$g_p^\dagger = \phi_{s^\dagger(p)}^u$$

Hence, for each $p \in \mathbb{N}$, we have:

- $p \in \mathbb{K}^u$ implies $\phi^u_{s^\dagger(p)} = f$;

- $p \notin \mathbb{K}^u$ implies $\phi^u_{s^\dagger(p)} = \lambda x.\,\text{undefined}$.

Therefore, given a computable enumeration $h^\dagger$ of $\mathbb{K}^u$,

$$\text{range}(s^\dagger \circ h^\dagger) \subseteq \text{ind}^u(f).$$

Moreover,

$$p \in \mathbb{K}^u \quad \text{iff} \quad s^\dagger(p) \in \text{range}(s^\dagger \circ h^\dagger).$$

Indeed:

- if $p \in \mathbb{K}^u$ then, since $h^\dagger$ is an enumeration of $\mathbb{K}^u$,

$$s^\dagger(p) \in s^\dagger(\mathbb{K}^u) = s^\dagger(\text{range}\, h^\dagger) = \text{range}(s^\dagger \circ h^\dagger);$$

- if $p \notin \mathbb{K}^u$ then $\phi^u_{s^\dagger(p)} = \lambda x.\,\text{undefined}$, from where we conclude that $s^\dagger(p) \notin \text{ind}^u(f)$, and, so,

$$s^\dagger(p) \notin \text{range}(s^\dagger \circ h^\dagger),$$

since, as already mentioned, range $(s^\dagger \circ h^\dagger) \subseteq \text{ind}^u(f)$.

Finally, we note that range$(s^\dagger \circ h^\dagger)$ is infinite. In fact, otherwise, range$(s^\dagger \circ h^\dagger)$ would be decidable and so, by Proposition 2.14, $\mathbb{K}^u$ would also be decidable, in contradiction with Proposition 3.6.

(2) $f = \lambda x.\,\text{undefined}$:

Consider now the function

$$g^\ddagger = \lambda\, p, x. \begin{cases} f(x) & \text{if } p \in P^u \\ 0 & \text{if } p \in Q^u \\ \text{undefined} & \text{otherwise} \end{cases} : \mathbb{N}^2 \rightharpoonup \mathbb{N}.$$

Recall Proposition 3.16. Function $g^\ddagger$ is well defined because $P^u \cap Q^u = \emptyset$. Moreover, $g^\ddagger$ is computable thanks to the listability of P^u and Q^u. For instance, $g^\ddagger$ is computed by the following program:

```
function (p,x) (
    t = 0;
    b₁ = -1;
    b₂ = -1;
    while ¬isnat(b₁) ∧ ¬isnat(b₂) do (
        t = t+1;
        b₁ = stceval(P_{χ^p_{pu}}, p, t, -1);
        b₂ = stceval(P_{χ^p_{Qu}}, p, t, -1);
    );
    if b₁ == 1 then
        return P_f(x)
    else
        return 0
)
```

where $P_{\chi^p_{pu}}$, $P_{\chi^p_{Qu}}$ and P_f are programs that compute χ^p_{Pu}, χ^p_{Qu} and f, respectively. On the other hand, for each $p \in \mathbb{N}$, we have:

- $p \in P^u$ implies $g^{\ddagger}_p = f$;

- $p \in Q^u$ implies $g^{\ddagger}_p = \lambda x.0$;

- $p \notin (P^u \cup Q^u)$ implies $g^{\ddagger}_p = \lambda x.\text{undefined}$.

Furthermore, since u is a proper universal function, there is $s^{\ddagger} \in m\mathscr{C}_1$ such that:

$$g^{\ddagger}_p = \phi^u_{s^{\ddagger}(p)}$$

for every $p \in \mathbb{N}$. Thus, given a computable enumeration $h^{\ddagger}$ of P^u,

$$\text{range}(s^{\ddagger} \circ h^{\ddagger}) \subseteq \text{ind}^u(f).$$

Furthermore, for each $p \in \mathbb{N}$, we have:

- $p \in Q^u$ implies $\phi^u_{s^{\ddagger}(p)} = \lambda x.0$;

- $p \notin Q^u$ implies $\phi^u_{s^{\ddagger}(p)} = f = \lambda x.\text{undefined}$.

Consider the set

$$C = \{p \in \mathbb{N} : s^{\ddagger}(p) \in \text{range}(s^{\ddagger} \circ h^{\ddagger})\} = (s^{\ddagger})^{-1}(\text{range}(s^{\ddagger} \circ h^{\ddagger})).$$

It is immediate to see that

$$p \in C \quad \text{iff} \quad s^{\ddagger}(p) \in \text{range}(s^{\ddagger} \circ h^{\ddagger})$$

for each $p \in \mathbb{N}$. We now show that C separates P^u and Q^u:

- $P^u \subseteq C$, since if $p \in P^u$ then $p \in \text{range}\, h^{\ddagger}$ and, so,

$$s^{\ddagger}(p) \in s^{\ddagger}(\text{range}\, h^{\ddagger}) = \text{range}(s^{\ddagger} \circ h^{\ddagger}).$$

- $C \cap Q^u = \emptyset$, since if $p \in C$ then $s^{\ddagger}(p) \in \text{range}(s^{\ddagger} \circ h^{\ddagger})$ and, so,

$$s^{\ddagger}(p) \in \text{ind}^u(f),$$

from where we obtain

$$s^{\ddagger}(p) \notin \text{ind}^u(\lambda x.0)$$

and, thus, $p \notin Q^u$.

Finally we show that $\text{range}(s^{\ddagger} \circ h^{\ddagger})$ is infinite. In fact, otherwise, $\text{range}(s^{\ddagger} \circ h^{\ddagger})$ would be decidable and so, by Proposition 2.14, C would also be decidable, contradicting Proposition 3.16.

Construction of the envisaged injective map:

Given $f \in \mathscr{C}_1$, in each case ($f \neq \lambda x.\text{undefined}$ and $f = \lambda x.\text{undefined}$), we were able to find computable maps $h : \mathbb{N} \to \mathbb{N}$ and $s : \mathbb{N} \to \mathbb{N}$ such that:

- $\text{range}(s \circ h) \subseteq \text{ind}^u(f)$;

- $\text{range}(s \circ h)$ is infinite.

Taking also into account that $\text{range}(s \circ h)$ is listable, by Proposition 2.11, we can conclude, by Proposition 2.9, that there is an injective computable enumeration $r : \mathbb{N} \to \mathbb{N}$ of $\text{range}(s \circ h)$ over $\mathbb{N}$. Then, r fulfills the requirements of the thesis: $r \in \text{m}\mathscr{C}_1$, r is injective and $\text{range}\, r \subseteq \text{ind}^u(f)$. $\qquad\square$

The following exercise shows that there is a computable binary map c where each section c_p is an injective computable enumeration of an infinite subset of the indices of the function with u-index p.

Exercise 4.8. Let u be a proper universal function. Show that there is $c \in \text{m}\mathscr{C}_2$ such that, for each $p \in \mathbb{N}$, the following conditions hold:

- c_p is injective;

- $\text{range}\, c_p \subseteq \text{ind}^u(\phi_p^u)$.

4.3 Rice-Shapiro-McNaughton-Myhill Theorem

Capitalizing on the Rice-Shapiro Theorem, the objective now is to establish, with respect to a proper universal function, a necessary and sufficient condition for the listability of the set of indices of a class of computable functions. To this end, we need first to provide a computable universal function for $\text{f}\mathscr{C}_1$. Such a function is presented in the proof of Proposition 4.7 below, using the map introduced in the following exercise.

Exercise 4.9. Present an algorithm for computing the map

$$b : \mathbb{N}^+ \times \mathbb{N} \to \bigcup_{m \in \mathbb{N}^+} \{\langle k_1, \ldots, k_m \rangle : k_j \in \mathbb{N}, j = 1, \ldots, m\}$$

such that

$$b(m, x) = \langle \beta_1^m(x) \rangle$$

for every $m, x \in \mathbb{N}$.

Extending a bit the terminology, we might say that map b is universal for the class $\{\beta_1^m : m \in \mathbb{N}^+\}$.

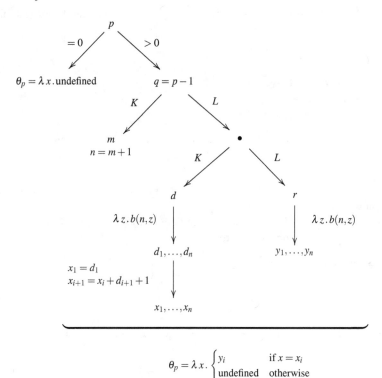

$$\theta_p = \lambda x . \begin{cases} y_i & \text{if } x = x_i \\ \text{undefined} & \text{otherwise} \end{cases}$$

Figure 4.1: Obtaining θ_p from $p \in \mathbb{N}$.

Proposition 4.7. There exists $\theta \in \mathscr{C}_2$ such that:

- $\{\theta_p : p \in \mathbb{N}\} = f\mathscr{C}_1$;

- $\theta_{p_1} = \theta_{p_2}$ implies $p_1 = p_2$, for every $p_1, p_2 \in \mathbb{N}$.

Proof. The idea for defining θ is depicted in Figure 4.1. That is, for each $p \in \mathbb{N}$, θ_p is such that

- θ_0 is the undefined unary function.

- for each $p > 0$, θ_p is the function such that $\operatorname{dom} \theta_p = \{x_1, \ldots, x_n\}$, range $\theta_p = \{y_1, \ldots, y_n\}$ and

$$\theta_p(x_i) = y_i, \text{ for } i = 1, \ldots, n$$

where $n > 0$ and $x_1 < \cdots < x_n$ and $y_1 \ldots, y_n$ are computed from p as follows:

- $(m, d, r) = b(3, q)$ where $q = p - 1$;

- $n = m + 1$;

- $(d_1, \ldots, d_n) = b(n, d)$;

- $(x_1, \ldots, \underline{x_i}, \ldots, x_n) = (d_1, \ldots, \underline{x_{i-1} + d_i + 1}, \ldots, x_{n-1} + d_n + 1)$;

- $(y_1, \ldots, y_n) = b(n, r)$.

We show that:

(1) $\{\theta_p : p \in \mathbb{N}\} = \mathsf{f}\mathscr{C}_1$. It is easy to see that each θ_p is a finite function. Moreover, θ_p is a computable function by Problem 1.1.

(2) Assume that $\vartheta \in \mathsf{f}\mathscr{C}_1$. Let $\operatorname{dom} \vartheta = \{x_1, \ldots, x_n\}$. Take

$$p = \beta_3^1(n - 1, d, r) + 1$$

where $d = \beta_n^1(x_1, \ldots, x_i - x_{i-1} - 1, \ldots, x_n - x_{n-1} - 1)$ and $r = \beta_n^1(\vartheta(x_1), \ldots, \vartheta(x_n))$. Then, θ_p is ϑ.

(3) Assume that $\theta_{p_1} = \theta_{p_2}$. Then, $\operatorname{dom} \theta_{p_1} = \operatorname{dom} \theta_{p_2} = \{x_1, \ldots, x_n\}$ with $x_1 < \cdots < x_n$ for some $n \in \mathbb{N}$. Moreover, $\theta_{p_1}(x_i) = \theta_{p_2}(x_i)$ for every $i = 1, \ldots, n$. Let

$$d = \beta_n^1(x_1, \ldots, x_i - x_{i-1} - 1, \ldots, x_n - x_{n-1} - 1) \quad \text{and} \quad r = \beta_n^1(\theta_{p_1}(x_1), \ldots, \theta_{p_1}(x_n)).$$

Then, by definition, $p_1 = \beta_3^1(n - 1, d, r) + 1 = p_2$.

(4) θ is a computable function. Let P_b be an algorithm that computes b. We start by introducing two programs. The first one $\mathsf{P}_{\mathsf{dom}}$ is as follows:

```
function (p) (
    if p == 0 then
        return ⟨⟩;
    n = P_b(3, p − 1)[1] + 1;
    d = P_b(3, p − 1)[2];
    w = P_b(n, d);
    i = 2;
    while i ≤ n do (
        w[i] = w[i − 1] + w[i] + 1;
        i = i + 1
    );
    return w
)
```

This program when executed on a natural number p returns the empty list when p is 0 and returns the list w with the elements in the domain of θ_p otherwise. That is, the number of elements n in the domain of θ_p is given by the first component of $b(3, p − 1)$ plus 1. Then we use the second component d of $b(3, p − 1)$ to generate, using b and n, a list with n elements that are used to obtain the elements in the domain in such a way that all of them are different. The second program P_{res} is

```
function (p) (
    if p == 0 then
        return ⟨⟩;
    n = P_b(3, p − 1)[1] + 1;
    r = P_b(3, p − 1)[3];
    return P_b(n, r)
)
```

This program when executed on a natural number p returns the empty list when p is 0 and returns the list w with the elements in the range of θ_p otherwise. That is, the number of elements n in the range of θ_p is given by the first component of $b(3, p − 1)$ plus 1. Then we use the third component of $b(3, p − 1)$ to generate, using b and n, the list of results.

Then, the program P_θ defined by

```
function (p, x) (
    k = pos(P_dom(p), x);
    if k == 0 then
        while true do null
    else
        return P_res(p)[k]
)
```

computes θ.   □

Bringing to finite functions the index terminology introduced before for computable functions, given $\vartheta \in f\mathscr{C}_1$, when $\theta_p = \vartheta$ we may say that p is the (unique) θ-*index* of the finite ϑ.

Proposition 4.8 (Rice-Shapiro-McNaughton-Myhill Theorem). Let u be a proper universal function and $A \subseteq \mathscr{C}_1$. Then:

$$\mathrm{ind}^u(A) \in \mathscr{L}_1 \quad \text{iff} \quad \exists P \in \mathscr{L}_1 : A = \bigcup_{p \in P} \{f \in \mathscr{C}_1 : \theta_p \subseteq f\}.$$

Proof. ($\rightarrow$) Assume that $\mathrm{ind}^u(A)$ is listable. Take

$$P = \{p \in \mathbb{N} : \theta_p \in A\}.$$

Then, as envisaged:

(1) P is listable. Indeed, since u is proper universal and $\theta \in \mathscr{C}_2$ by Propostion 4.7, there exists $s \in m\mathscr{C}_1$ such that

$$\theta_p = \phi_{s(p)}^u$$

for every $p \in \mathbb{N}$. Thus

$$P = \{p \in \mathbb{N} : \phi_{s(p)}^u \in A\}.$$

That is,

$$p \in P \quad \text{iff} \quad s(p) \in \mathrm{ind}^u(A).$$

Hence, P is listable because

$$\chi_P^p = \chi_{\mathrm{ind}^u(A)}^p \circ s$$

is computable thanks to the listability of $\mathrm{ind}^u(A)$.

(2) $A \subseteq \bigcup_{p \in P} \{f \in \mathscr{C}_1 : \theta_p \subseteq f\}$:

Let $g \in A$. Then, thanks to the listability of $\mathrm{ind}^u(A)$, by the Rice-Shapiro Theorem, there is finite $\vartheta \in A$ such that $\vartheta \subseteq g$. Using Proposition 4.7, there is $q \in \mathbb{N}$ such that $\theta_q = \vartheta$. That is:

- $q \in P$ (by choice of P, since $\theta_q \in A$);

- $\theta_q \subseteq g$.

Thus, $g \in \bigcup_{p \in P} \{f \in \mathscr{C}_1 : \theta_p \subseteq f\}$.

(3) $A \supseteq \bigcup_{p \in P} \{f \in \mathscr{C}_1 : \theta_p \subseteq f\}$. Let $g \in \bigcup_{p \in P} \{f \in \mathscr{C}_1 : \theta_p \subseteq f\}$. Then, there is $q \in P$ such that:

- $\theta_q \subseteq g$;

- $\theta_q \in A$ (by choice of P, since $q \in P$).

Hence, thanks to the listability of $\text{ind}^u(A)$, by the Rice-Shapiro Theorem, $g \in A$.

($\leftarrow$) Assume that

$$A = \bigcup_{p \in P} \{f \in \mathscr{C}_1 : \theta_p \subseteq f\}$$

for some $P \in \mathscr{L}_1$. We have to show that $\text{ind}^u(A) \in \mathscr{L}_1$. To this end, consider the set

$$C = \{(p,q) \in \mathbb{N}^2 : \theta_p \subseteq \phi_q^u\}.$$

We show that C is listable by computing its characteristic function χ_C^P as follows:

```
function (p,q) (
    w = P_dom(p);
    j = 1;
    while j ≤ length(w) do (
        if P_θ(p,w[j]) ≠ P_u(q,w[j]) then
            while true do null;
        j = j+1
    );
    return 1
)
```

where P_{dom} and P_θ are defined in the proof of Proposition 4.7 and P_u is a program that computes u. Observe that this procedure terminates (in which case it returns 1) if and only if ϕ_q^u and θ_p coincide on all the elements in the domain of θ_p, that is, if and only if $\theta_p \subseteq \phi_q^u$. Furthermore, as required,

$$
\begin{aligned}
\text{ind}^u(A) &= \text{ind}^u\left(\bigcup_{p \in P} \{f \in \mathscr{C}_1 : \theta_p \subseteq f\}\right) \\[2mm]
&= \bigcup_{p \in P} \text{ind}^u(\{f \in \mathscr{C}_1 : \theta_p \subseteq f\}) \\[2mm]
&= \bigcup_{p \in P} \{q \in \mathbb{N} : \theta_p \subseteq \phi_q^u\} \\[2mm]
&= \{q \in \mathbb{N} : \exists\, p \in P \text{ such that } \theta_p \subseteq \phi_q^u\} \\[2mm]
&= \{q \in \mathbb{N} : \exists\, p \in P \text{ such that } (p,q) \in C\} \\[2mm]
&= \{q \in \mathbb{N} : \exists\, p \in \mathbb{N} \text{ such that } (p,q) \in (P \times \mathbb{N}) \text{ and } (p,q) \in C\} \\[2mm]
&= \text{proj}_{[2]}^{\mathbb{N}^2}((P \times \mathbb{N}) \cap C)
\end{aligned}
$$

is listable, since C and P are both listable and, by Proposition 2.7, $\mathrm{proj}^{\mathbb{N}^2}_{[2]}$ preserves listability. $\square$

Example 4.2. We now show that the set

$$\{p \in \mathbb{N} : \phi_p(0) = 0\}$$

is listable. Observe that

$$\{p \in \mathbb{N} : \phi_p(0) = 0\} = \mathrm{ind}(\{f \in \mathscr{C}_1 : f(0) = 0\}).$$

Let $P = \{p_0\}$ where p_0 is the unique θ-index such that

$$\theta_{p_0} = \lambda x . \begin{cases} 0 & \text{if } x = 0 \\ \text{undefined} & \text{otherwise} \end{cases}.$$

Observe that P is listable since it is decidable by Exercise 2.12. It is immediate to see that

$$\{f \in \mathscr{C}_1 : f(0) = 0\} = \bigcup_{p \in P} \{f \in \mathscr{C}_1 : \theta_p \subseteq f\}.$$

Hence, by the Rice-Shapiro-McNaughton-Myhill Theorem, the set $\{p \in \mathbb{N} : \phi_p(0) = 0\}$ is listable.

Exercise 4.10. Show that the set $\mathrm{ind}(\{f \in \mathscr{C}_1 : \exists k \in \mathbb{N} \ f(k) = 0\})$ is listable.

With the Rice-Shapiro-McNaughton-Myhill Theorem at hand, it is finally possible to show that the RSC is not a sufficient condition. The reader is invited to present a counterexample in the following exercise.

Exercise 4.11. Let u be a proper universal function. Provide $A \subseteq \mathscr{C}_1$ such that A fulfills the Rice-Shapiro condition but $\mathrm{ind}^u(A)$ is non-listable.

4.4   Rogers' Theorem

The fact that an injective computable enumeration of a subset of the set of indices of any computable function can be effectively obtained from an index of the given function (Exercise 4.8) is the key ingredient for proving that any two proper universal functions are isomorphic (a result attributed to Hartley Rogers Jr., [52]). To this end, it is worthwhile to set up the relevant category of binary functions (for more details on categories see [39]).

Definition 4.2. Given $v, v' \in \mathscr{F}_2$, by a *(binary-function) section-translation morphism*

$$s : v \to v'$$

we understand a computable map $s : \mathbb{N} \to \mathbb{N}$ such that

$$v_p = v'_{s(p)}$$

for every $p \in \mathbb{N}$.

That is, a morphism $s : v \to v'$ allows the recovery of each section of v as a section of v'.

Definition 4.3. Given $s : v \to v'$ and $s' : v' \to v''$, the map $s' \circ s : \mathbb{N} \to \mathbb{N}$ establishes a morphism $s' \circ s : v \to v''$, called their *composition*.

Exercise 4.12. Show that composition of morphisms is associative.

Exercise 4.13. Let $v \in \mathscr{F}_2$. Show that $id_{\mathbb{N}}$ is a section-translation morphism from v to v (the *identity morphism* for v), denoted by

$$id_v$$

such that

$$s \circ id_v = s \quad \text{and} \quad id_{v'} \circ s = s$$

for each $s : v \to v'$.

Remark 4.3. We denote by $\mathbf{F}_2$ the category of binary functions and their section-translation morphisms. Moreover, we denote by $\mathbf{C}_2$ the full subcategory of $\mathbf{F}_2$ composed of all computable binary functions and all their section-translation morphisms.

Our objective now is to provide a categorical characterization of proper universal functions. Before we introduce an auxiliary result.

Proposition 4.9. A computable binary function v is universal if there is $s : \text{univ} \to v$.

Proof. Since v is computable,

$$\{v_p : p \in \mathbb{N}\} \subseteq \mathscr{C}_1,$$

by Exercise 3.1. Moreover, by Proposition 3.2 and using morphism $s : \text{univ} \to v$,

$$\mathscr{C}_1 = \{\text{univ}_p : p \in \mathbb{N}\} = \{v_{s(p)} : p \in \mathbb{N}\} \subseteq \{v_p : p \in \mathbb{N}\}.$$

Therefore, v is universal.   □

Proposition 4.10. A computable binary function v is proper universal if and only if for each computable binary function v' there is a morphism $s : v' \to v$.

Proof. The existence of a morphism $s : v' \to v$ for each $v' \in \mathscr{C}_2$ is just a restatement, in the categorical language, of asserting that v enjoys the *s-m-n* property. Recall also that every computable binary function enjoying the *s-m-n* property is proper universal.   □

Proposition 4.11. A computable binary function v is proper universal if and only if there is a morphism $s : \text{univ} \to v$.

Proof. ($\to$) Assume that $v \in \mathscr{C}_2$ is a proper universal function. Then, by Proposition 4.10, there is a morphism $s : \text{univ} \to v$, since $\text{univ} \in \mathscr{C}_2$.

($\leftarrow$) Assume that there is a morphism $s : \text{univ} \to v$. Let $v' \in \mathscr{C}_2$. Since univ is a proper universal function then, by Proposition 4.10, there is morphism $s : v' \to \text{univ}$. Hence, $s \circ s' : v' \to v$ is a morphism and so, again, by Proposition 4.10, v is a proper universal function. □

Definition 4.4. A binary function $v \in \mathbf{C}_2$ is said to be *protofinal* in $\mathbf{C}_2$ if there is a section-translation morphism from each $v' \in \mathbf{C}_2$ to v. Moreover, v is said to be *final* in $\mathbf{C}_2$ if each such morphism is unique.

Therefore, the proper universal functions are the protofinal objects in $\mathbf{C}_2$, by Proposition 4.10. They are not quite final because the uniqueness requirement is not fulfilled.

Proposition 4.12. There are no final objects in $\mathbf{C}_2$.

Proof. Assume that u is final in $\mathbf{C}_2$. Then, for every $v \in \mathscr{C}_2$, there is a unique morphism $s : v \to u$. Hence, by Proposition 4.10, u is a proper universal function. Moreover, there is a unique morphism from u to u. Since $\text{id}_u : u \to u$, then $\text{id}_{\mathbb{N}} : \mathbb{N} \to \mathbb{N}$ is the unique permutation of $\mathbb{N}$ which preserves the sections of u. Therefore, for any given $f \in \mathscr{C}_1$,

$$|\text{ind}^u(f)| = 1.$$

Thus, $\text{ind}^u(f)$ is a decidable set in contradiction with Rice's theorem that states that the only decidable sets of indices are those corresponding to either $\emptyset$ or to $\mathscr{C}_1$. □

Definition 4.5. A morphism $s : v \to v'$ in $\mathbf{C}_2$ is said to be an *isomorphism* in $\mathbf{C}_2$ if there is a morphism $s' : v' \to v$ in $\mathbf{C}_2$ such that

$$s' \circ s = \text{id}_v \text{ and } s \circ s' = \text{id}_{v'}.$$

In this case s' is said to be an *inverse* of s in $\mathbf{C}_2$.

Exercise 4.14. Show that whenever a morphism has an inverse then it is unique.

We denote by s^{-1} the inverse (when it exists) of a morphism s.

Exercise 4.15. Show that the inverse of an isomorphism exists and is also an isomorphism.

Exercise 4.16. Show that s is an isomorphism if and only if $s : \mathbb{N} \to \mathbb{N}$ is a bijection.

Proposition 4.13 (Rogers' Theorem). Any two proper universal functions are isomorphic in $\mathbf{C}_2$.

Proof. Let u and u' be proper universal functions. Taking into account Exercise 4.8, let $c, c' : \mathbb{N}^2 \to \mathbb{N}$ be computable maps such that, for each $p \in \mathbb{N}$, the following conditions hold:

- c_p and c'_p are injective;

- range $c_p \subseteq \mathrm{ind}^u(\phi_p^u)$ and range $c'_p \subseteq \mathrm{ind}^{u'}(\phi_p^{u'})$.

Moreover, by Proposition 4.10, let $r : u \to u'$ and $r' : u' \to u$ be morphisms in $\mathbf{C}_2$. We want to define an injective and computable map

$$h : \mathbb{N} \to \mathbb{N}^2$$

such that

$$\mathrm{range}\,(\mathrm{proj}_{[1]}^{\mathbb{N}^2} \circ h) = \mathrm{range}\,(\mathrm{proj}_{[2]}^{\mathbb{N}^2} \circ h) = \mathbb{N}$$

and for each $k \in \mathbb{N}$,

- $\phi^u_{(\mathrm{proj}_{[1]}^{\mathbb{N}^2} \circ h)(k)} = \phi^{u'}_{(\mathrm{proj}_{[2]}^{\mathbb{N}^2} \circ h)(k)}$;

- $\{h(0), \dots, h(k)\}$ is the graph of a bijection from $\{(\mathrm{proj}_{[1]}^{\mathbb{N}^2} \circ h)(0), \dots, (\mathrm{proj}_{[1]}^{\mathbb{N}^2} \circ h)(k)\}$ to $\{(\mathrm{proj}_{[2]}^{\mathbb{N}^2} \circ h)(0), \dots, (\mathrm{proj}_{[2]}^{\mathbb{N}^2} \circ h)(k)\}$.

Let P_c, $P_{c'}$, P_r and $P_{r'}$ be programs that compute c, c', r and r', respectively, P_{count} as defined in Example 1.13 and let P_σ be the program

```
function (w) (
    i = 0;
    while P_count(w, i) ≠ 0 do
        i = i + 1;
    return i
)
```

Observe that P_σ when executed on a list w returns the first natural number not in w. Let h be the map computed by the following algorithm P_h:

```
function (k) (
    w1, w2 = ⟨⟩, ⟨⟩;
    p = 0;
    while p ≤ k do (
        if p%2 == 0 then (
            p1 = Pσ(w1);
            j = 0;
            p2 = Pc'(Pr(p1), j);
            while Pcount(w2, p2) ≠ 0 do (
                j = j + 1;
                p2 = Pc'(Pr(p1), j)
            )
        )
        else (
            p2 = Pσ(w2);
            j = 0;
            p1 = Pc(Pr'(p2), j);
            while Pcount(w1, p1) ≠ 0 do (
                j = j + 1;
                p1 = Pc(Pr'(p2), j)
            )
        );
        w1, w2 = append(w1, p1), append(w2, p2)
    );
    return w1[k], w2[k]
)
```

The idea behind P_h is that, given k, it generates an index over u' not previously used corresponding to the least index over u not used before and vice versa, and this process is repeated k-times. In the end, the execution of P_h on k returns the k-th indices of u and u'. Then

- $s = (\text{proj}_{[2]}^{\mathbb{N}^2} \circ h) \circ (\text{proj}_{[1]}^{\mathbb{N}^2} \circ h)^{-1} : \mathbb{N} \to \mathbb{N};$

- $s' = (\text{proj}_{[1]}^{\mathbb{N}^2} \circ h) \circ (\text{proj}_{[2]}^{\mathbb{N}^2} \circ h)^{-1} : \mathbb{N} \to \mathbb{N}.$

are isomorphisms in $\mathbf{C}_2$ between u and u' with $s' = s^{-1}$. That is,

$$u_{p_1} = u'_{s(p_1)} \quad \text{and} \quad u'_{p_2} = u_{s'(p_2)}$$

for each $p_1, p_2 \in \mathbb{N}$. Observe that, in this case, s and s' are computable maps by Proposition 1.5, Proposition 1.2 and Proposition 1.7 and since h is a computable map. So, h has the envisaged properties described above. Therefore, u and u' are isomorphic. □

Exercise 4.17. Consider the map h defined in the proof of Proposition 4.13 (Rogers' Theorem). Show that $\{h(0),\dots,h(k)\}$ is the graph of a bijection between the sets $\{h(0)_1,\dots,h(k)_1\}$ and $\{h(0)_2,\dots,h(k)_2\}$.

As the reader might expect, there is a counterpart of Rogers' Theorem for proper universal sets.

Definition 4.6. Given $V,V' \in \mathscr{S}_2$, by a *(binary-set) section-translation morphism*

$$s : V \to V'$$

we mean a computable map $s : \mathbb{N} \to \mathbb{N}$ such that

$$V_p = V'_{s(p)}$$

for every $p \in \mathbb{N}$.

Exercise 4.18. Establish the relevant category $\mathbf{S}_2$ of binary sets and their morphisms, as well as its full subcategory $\mathbf{L}_2$ of listable binary sets. Show that the isomorphisms are the bijections. Show that any two proper universal sets are isomorphic.

The counterpart of Rogers' Theorem for proper universal sets provides the means for proving the converse of Proposition 3.17.

Proposition 4.14. A binary set is proper universal if and only if it is the domain of a proper universal function.

Proof. ($\leftarrow$) Assume that the binary set U is the domain of a proper universal function. Then, by Proposition 3.17, U is a proper universal set.

($\to$) Let U be a proper universal set. Taking into account Exercise 4.18, let $s : U \to$ Univ be an isomorphism. Consider the binary computable function

$$u = \lambda\, p,x \,.\, \mathrm{univ}_{s(p)}(x).$$

We now show that u is a proper universal function. Let $v \in \mathscr{C}_2$. Then, there is $r \in \mathrm{m}\mathscr{C}_1$ such that

$$v_p = \mathrm{univ}_{r(p)}$$

for every $p \in \mathbb{N}$. So

$$v_p = \mathrm{univ}_{r(p)} = \mathrm{univ}_{s(s^{-1}(r(p)))} = u_{s^{-1}(r(p))}$$

for every $p \in \mathbb{N}$. Observe that $s^{-1} \circ r$ is a computable unary map since r is a computable map and s is a computable bijection. Hence, u is a proper universal function. Furthermore,

$$(p,x) \in U \text{ iff } (s(p),x) \in \mathrm{Univ} \text{ iff } (s(p),x) \in \mathrm{dom\,univ} \text{ iff } (p,x) \in \mathrm{dom}\,u$$

and so $U = \mathrm{dom}\,u$. $\square$

4.5   Recursion Theorem

The contents of this section have also significant practical consequences namely on answering questions related with the possibility of avoiding computer virus. As we shall see the existence of computer virus is a consequence of the *s-m-n* property. Nevertheless, the *s-m-n* property is an essential requirement for a viable computational model. We start with some preliminary notions and results.

Definition 4.7. Let $u \in \mathscr{F}_2$. Consider the binary relation on $\mathbb{N}$

$$\approx^u$$

defined as follows:

$$p_1 \approx^u p_2 \quad \text{whenever} \quad \phi_{p_1}^u = \phi_{p_2}^u$$

for each $p_1, p_2 \in \mathbb{N}$.

Exercise 4.19. Show that $\approx^u$ is an equivalence relation.

Definition 4.8. Let $\approx$ be an equivalence relation in $\mathbb{N}$. Given $f, g \in \mathscr{F}_1$ we say that g is an $\approx$-*extension* of f, written

$$f \precsim g,$$

if $g(x) \approx f(x)$ for each $x \in \operatorname{dom} f$.

Exercise 4.20. Show that $\precsim$ coincides with $\subseteq$ when $\approx$ is $=$.

Definition 4.9. We say that the equivalence relation $\approx$ in $\mathbb{N}$ enjoys property $\forall$CME whenever every computable unary function is $\approx$-extendable to a computable map.

Exercise 4.21. Show that the equivalence relation $=$ does not enjoy the $\forall$CME property.

Definition 4.10. Given $f \in \mathscr{C}_1$, we say that $x \in \mathbb{N}$ is an $\approx$-*fixed point* of f whenever $f(x) \approx x$.

Exercise 4.22. Identify the set of all $=$-fixed points for $\lambda x . x + 1$ and $id_{\mathbb{N}}$.

Definition 4.11. We say that the equivalence relation $\approx$ in $\mathbb{N}$ enjoys property $\exists$NFP whenever there is a computable map without any $\approx$-fixed point.

Exercise 4.23. Show that the equivalence relation $=$ enjoys the $\exists$NFP property.

Proposition 4.15. No equivalence relation in $\mathbb{N}$ can enjoy both $\forall$CME and $\exists$NFP properties.

Proof. Assume, by contradiction, that there is an equivalence relation $\approx$ in $\mathbb{N}$ enjoying both properties. Thanks to Proposition 3.2 and Proposition 3.4, let f be a computable function such that no computable function differs from it everywhere. By property $\forall$CME, there is a computable map g such that $f \precsim g$. By property $\exists$NFP, let h be a computable map without $\approx$-fixed points. Consider the map

$$e = h \circ g$$

which is obviously computable by Proposition 1.2. Observe that:

- if $x \in \text{dom} f$ then $f(x) \approx g(x) \not\approx h(g(x)) \approx e(x)$ and, so, $f(x) \neq e(x)$ since otherwise $f(x) \approx e(x)$ because $\approx$ is a reflexive relation;

- if $x \notin \text{dom} f$ then either f is undefined at x or $f(x) \notin \mathbb{N}$, while $e(x) \in \mathbb{N}$. Hence $f(x) \neq e(x)$.

Therefore, e differs from f everywhere contradicting the properties of f. □

We now show that $\approx^u$ does not have the $\exists$NFP property.

Proposition 4.16. Let u be a proper universal function and $f \in \mathscr{C}_1$. Then, $\approx^u$ enjoys the $\forall$CME property. Moreover, $\approx^u$ does not have the $\exists$NFP property.

Proof. Let $f \in \mathscr{C}_1$. Consider the computable binary function

$$v = \lambda p, x . u(f(p), x).$$

Since u is a proper universal function, there is $r \in \text{m}\mathscr{C}_1$ such that

$$u(f(p), x) = v(p, x) = u(r(p), x)$$

for every $p, x \in \mathbb{N}$. Thus, in particular,

$$p \in \text{dom} f \quad \text{implies} \quad \phi^u_{r(p)} = \phi^v_p = \phi^u_{f(p)}$$

for every $p \in \mathbb{N}$. Thus, $f \precsim^u r$. Therefore, $\approx^u$ has the $\forall$CME property and so, by Proposition 4.15, does not have the $\exists$NFP property. □

We are ready to discuss the first Recursion Theorem.

Proposition 4.17. [Recursion Theorem] Let u be a proper universal function and $s \in \text{m}\mathscr{C}_1$. Then, there exists $p \in \mathbb{N}$ such that

$$\phi^u_{s(p)} = \phi^u_p.$$

Proof. First, observe that, by Proposition 4.16, $\approx^u$ does not enjoy the $\exists$NFP property. That is, every computable map has an $\approx^u$-fixed point. In particular, s has an $\approx^u$-fixed point. Hence, there is $p \in \mathbb{N}$ such that

$$s(p) \approx^u p$$

and, so, there is $p \in \mathbb{N}$ such that $\phi^u_{s(p)} = \phi^u_p$. □

Exercise 4.24. State and prove the Recursion Theorem for proper universal sets.

Moreover, we show that every computable unary map has an infinite number of $\approx^u$-fixed points.

Proposition 4.18. Let u be a proper universal function, $s \in m\mathscr{C}_1$ and $q \in \mathbb{N}$. Then, there is $p \in \mathbb{N}$ such that

$$p > q \quad \text{and} \quad \phi^u_{s(p)} = \phi^u_p.$$

Proof. Let $c \in \mathbb{N}$ be such that:

$$\phi^u_c \neq \phi^u_0, \ldots, \phi^u_q.$$

Such a c exists. Indeed assume, by contradiction, that there is no such c. Then,

$$\{u_k : k \in \mathbb{N}\} = \{u_0, \ldots, u_q\}$$

is a finite set and so different from $\mathscr{C}_1$, contradicting the fact that u is a universal function. Consider the following map:

$$r = \lambda x. \begin{cases} c & \text{if } x \leq q \\ s(x) & \text{otherwise} \end{cases}.$$

Observe that if $k \leq q$ then $\phi^u_{r(k)} = \phi^u_c$ which is different from ϕ^u_k thanks to the choice of c. Hence,

$$\text{if } \phi^u_{r(k)} = \phi^u_k \text{ then } k > q$$

for every $k \in \mathbb{N}$. On the other hand, since r is computable, by the Recursion Theorem (Proposition 4.17), there is $p \in \mathbb{N}$ such that

$$\phi^u_{r(p)} = \phi^u_p.$$

Hence, $p > q$ and $s(p) = r(p)$. Therefore, the thesis follows. □

The interpretation of Proposition 4.18 is as follows: it is possible to find a $\approx^u$-fixed point of a map s as big as we want. We are now ready to address the problem of avoiding computer virus. Intuitively, a computer virus is a program that can replicate itself. In the context of computable functions, given a particular enumeration of computable functions, we can ask if there is a function that can "replicate" itself, that is, if there is a function that on any given input always outputs its own index? If so, we call the index of this function a virus.

Definition 4.12. Let $u \in \mathscr{F}_2$. A *virus* for u is a natural number p such that $\phi_p^u = \lambda x . p$.

We now show that, for proper universal functions, viruses exist.

Proposition 4.19. Let u be a proper universal function. Then, there is a virus for u.

Proof. Consider the function

$$v = \lambda q, x . q.$$

Function v is clearly computable. As u is a proper universal function, then there is $s \in \mathrm{m}\mathscr{C}_1$ such that

$$v_q = u_{s(q)}$$

for every $q \in \mathbb{N}$. Hence,

$$\phi_{s(q)}^u(x) = v(q, x) = q$$

for every $q, x \in \mathbb{N}$. On the other hand, by the Recursion Theorem (Propostion 4.17) there is $p \in \mathbb{N}$ such that

$$\phi_{s(p)}^u = \phi_p^u.$$

Thus, $\phi_p^u(x) = \phi_{s(p)}^u(x) = p$ for all $x \in \mathbb{N}$. □

The following result states a deep relationship between the listable set $\mathbb{K}$ and the Recursion Theorem.

Proposition 4.20. The undecidability of $\mathbb{K}$ is a corollary of the Recursion Theorem.

Proof. Recall that $\mathbb{K} = \mathrm{dom}\,(\lambda x . \mathrm{univ}(x, x))$ and $\mathrm{Univ} = \mathrm{dom}\,\mathrm{univ}$. Hence,

$$q \in \mathbb{K} \quad \text{iff} \quad q \in \mathrm{Univ}_q$$

for each $q \in \mathbb{N}$. Choose $a, b \in \mathbb{N}$ such that $\mathrm{Univ}_a = \emptyset$ and $\mathrm{Univ}_b = \mathbb{N}$. These indices exist since $\{\mathrm{Univ}_p : p \in \mathbb{N}\}$ is the set of all unary listable sets by Proposition 3.18. Towards a contradiction, assume that $\mathbb{K}$ is decidable. Then, the map

$$s = \lambda x . \begin{cases} a & \text{if } x \in \mathbb{K} \\ b & \text{otherwise} \end{cases}$$

is computable and, so, by the Recursion Theorem for proper universal sets (see Exercise 4.24) there is $p \in \mathbb{N}$ such that

$$(\dagger) \qquad \mathrm{Univ}_{s(p)} = \Gamma_{s(p)} = \Gamma_p = \mathrm{Univ}_p.$$

On the other hand, by the choice of s, we have:

$$\begin{cases} q \in \mathrm{Univ}_{s(q)} \;\Rightarrow\; \mathrm{Univ}_{s(q)} \neq \emptyset \;\Rightarrow\; s(q) = b \;\Rightarrow\; q \notin \mathbb{K} \;\Rightarrow\; q \notin \mathrm{Univ}_q \\ q \notin \mathrm{Univ}_{s(q)} \;\Rightarrow\; \mathrm{Univ}_{s(q)} \neq \mathbb{N} \;\Rightarrow\; s(q) = a \;\Rightarrow\; q \in \mathbb{K} \;\Rightarrow\; q \in \mathrm{Univ}_q \end{cases}$$

for each $q \in \mathbb{N}$. So, $\mathrm{Univ}_{s(q)} \neq \mathrm{Univ}_q$ for each $q \in \mathbb{N}$, in contradiction with $(\dagger)$. □

Observe that the recursion results proved above are not constructive in the sense that they do not provide a clue on how to find an $\approx^u$-fixed point of a given computable map. In fact, we can do better. The following result shows that there is an algorithm that computes an $\approx^u$-fixed point of a map when provided with one of the map indices as input. However, the algorithm when applied to an index of a non-total function may return a point outside the domain of that function.

Proposition 4.21 (Constructive Recursion Theorem)**.** Let u be a proper universal function. Then, there is $\varphi \in m\mathscr{C}_1$ such that

$$\begin{cases} \varphi(p) \in \operatorname{dom}\phi_p^u & \text{implies} & \phi_{\phi_p^u(\varphi(p))}^u = \phi_{\varphi(p)}^u \\ \varphi(p) \notin \operatorname{dom}\phi_p^u & \text{implies} & \operatorname{dom}\phi_{\varphi(p)}^u = \emptyset \end{cases}$$

for every $p \in \mathbb{N}$.

Proof. We want to find a computable unary map φ that for each $p \in \mathbb{N}$ provides an $\approx^u$-fixed point of ϕ_p^u. Let

$$\varphi = r \circ r'$$

where r and r' are defined in (1) and (2) below, respectively.

(1) Definition of a computable unary map r such that $\operatorname{diag}^u \precsim^u r$.

By Proposition 4.16, since $\approx^u$ enjoys $\forall$CME, let $r \in m\mathscr{C}_1$ be such that

$$(\dagger) \qquad u(\phi_p^u(p),x) = u(r(p),x)$$

for every $p \in \operatorname{dom}\operatorname{diag}^u$ and $x \in \mathbb{N}$.

(2) Definition of a computable unary map r' such that $\phi_p^u \circ r = \phi_{r'(p)}^u$. Let

$$v' = \lambda\, p,x.\, u(p,r(x)).$$

Then v' is computable. Since u is a proper universal function, there is $r' \in m\mathscr{C}_1$ such that

$$\phi_p^u(r(x)) = u(p,r(x)) = v'(p,x) = u(r'(p),x) = \phi_{r'(p)}^u(x)$$

for every $p,x \in \mathbb{N}$, and, so,

$$(\dagger\dagger) \qquad \phi_p^u \circ r = \phi_{r'(p)}^u$$

for each $p \in \mathbb{N}$.

Observe that

$$\begin{array}{lll} r'(p) \in \operatorname{dom}\operatorname{diag}^u & \text{iff} & \phi_{r'(p)}^u(r'(p))\!\downarrow \\ & \text{iff} & (\phi_p^u \circ r)(r'(p))\!\downarrow \\ & \text{iff} & r(r'(p)) \in \operatorname{dom}\phi_p^u. \end{array}$$

Then, for every $p, x \in \mathbb{N}$ such that $\varphi(p) \in \mathrm{dom}\, \phi_p^u$,

$$
\begin{aligned}
u(\phi_p^u(\varphi(p)), x) &= u((\phi_p^u \circ \varphi)(p), x) \\
&= u((\phi_p^u \circ (r \circ r'))(p), x) \\
&= u(((\phi_p^u \circ r) \circ r')(p), x) \\
&= u((\phi_{r'(p)}^u \circ r')(p), x) &(\dagger\dagger) \\
&= u(\phi_{r'(p)}^u(r'(p)), x) \\
&= u(r(r'(p)), x) &(\dagger) \\
&= u((r \circ r')(p), x) \\
&= u(\varphi(p), x).
\end{aligned}
$$

Hence, if $\varphi(p) \in \mathrm{dom}\, \phi_p^u$ then

$$
\phi_{\phi_p^u(\varphi(p))}^u = \phi_{\varphi(p)}^u.
$$

Moreover, if $\varphi(p) \notin \mathrm{dom}\, \phi_p^u$ then

$$
\phi_{\varphi(p)}^u(x) = u(\phi_p^u(\varphi(p)), x) = \text{undefined}
$$

for every $x \in \mathbb{N}$, and, so, $\phi_{\varphi(p)}^u$ is the undefined function. □

It is also worth mentioning that the Recursion Theorem (Proposition 4.17) is a corollary of the Constructive Recursion Theorem (Proposition 4.21). The following result is a parameterized version of the latter for maps.

Proposition 4.22. Let u be a proper universal function and $h \in \mathrm{m}\mathscr{C}_2$. Then, there is $\varphi \in \mathrm{m}\mathscr{C}_1$ such that

$$
\phi_{h(p, \varphi(p))}^u = \phi_{\varphi(p)}^u
$$

for every $p \in \mathbb{N}$.

Proof. The proof is similar to the proof of Proposition 4.21: it is enough to replace u by h in the definition of v'. □

The following counterpart of Proposition 4.22 for indexing listable sets will play a key role in the proof of Myhill's Theorem (Proposition 5.26).

Proposition 4.23. Let U be a proper universal set and $h \in \mathrm{m}\mathscr{C}_2$. Then, there is $\varphi \in \mathrm{m}\mathscr{C}_1$ such that

$$
\Gamma_{h(p, \varphi(p))}^U = \Gamma_{\varphi(p)}^U
$$

for every $p \in \mathbb{N}$.

Proof. Taking into account Proposition 4.14, there is a proper universal function u such that $U = \operatorname{dom} u$ whenever U is a proper universal set. Observe that, by Proposition 4.22,

$$u_{h(p,\varphi(p))} = \phi^u_{h(p,\varphi(p))} = \phi^u_{\varphi(p)} = u_{\varphi(p)}$$

for every $p \in \mathbb{N}$. Thus,

$$
\begin{aligned}
\Gamma^{\operatorname{dom} u}_{h(p,\varphi(p))} &= (\operatorname{dom} u)_{h(p,\varphi(p))} \\
&= \operatorname{dom} u_{h(p,\varphi(p))} \\
&= \operatorname{dom} u_{\varphi(p)} \\
&= (\operatorname{dom} u)_{\varphi(p)} \\
&= \Gamma^{\operatorname{dom} u}_{\varphi(p)}
\end{aligned}
$$

for every $p \in \mathbb{N}$, as envisaged.   □

4.6   Solved Problems and Exercises

Problem 4.1. Let u be a proper universal function and $k \in \mathbb{N}$. Show that the problem

$$\text{Given } p \in \mathbb{N}, \text{ is } \phi^u_p = \lambda x . k?$$

is non-decidable.

Solution We use the reduction technique, introduced in Section 2.3, to prove that the given problem is non-decidable. Consider the set

$$A = \{\lambda x . k\}.$$

Observe that

$$
\begin{aligned}
p \in \operatorname{ind}^u(A) \qquad &\text{iff} \qquad p \in \{q \in \mathbb{N} : \phi^u_q \in A\} \\
&\text{iff} \qquad p \in \{q \in \mathbb{N} : \phi^u_q = \lambda x . k\} \\
&\text{iff} \qquad \phi^u_p = \lambda x . k.
\end{aligned}
$$

Therefore the problem

$$\text{Given } p \in \mathbb{N}, \text{ does } p \in \operatorname{ind}^u(A)?$$

can be reduced to problem

$$\text{Given } p \in \mathbb{N}, \text{ is } \phi^u_p = \lambda x . k?$$

Hence, the decidability of the latter would entail the decidability of $\operatorname{ind}^u(A)$ (see Proposition 2.14). But, as $\emptyset \subsetneq A \subsetneq \mathscr{C}_1$, by Rice's Theorem, Proposition 4.4, $\operatorname{ind}^u(A)$ is non-decidable. Thus, the given problem is non-decidable.   ▽

Problem 4.2. Let u be a proper universal function and $k \in \mathbb{N}$. Show that the problem

$$\text{Given } p \in \mathbb{N}, \text{ does } k \in \text{range } \phi_p^u?$$

is non-decidable, without using Rice's Theorem (Proposition 4.4).

Solution We present a direct proof, without using Rice's Theorem. We show that the problem

$$\text{Given } p \in \mathbb{N}, \text{ does } p \in \mathbb{K}^u?$$

can be reduced to the problem

$$\text{Given } p \in \mathbb{N}, \text{ does } p \in \{q \in \mathbb{N} : k \in \text{range } \phi_q^u\}?$$

Consider the following binary function

$$v = \lambda\, p,x\,.\begin{cases} x & \text{if } p \in \mathbb{K}^u \\ \text{undefined} & \text{otherwise.} \end{cases}$$

Observe that

$$v = (\lambda\, x,y\,.\,x \times y) \circ \langle \chi_{\mathbb{K}^u}^{\mathbb{P}} \circ \text{proj}_{[1]}^{\mathbb{N}^2}, \text{proj}_{[2]}^{\mathbb{N}^2} \rangle.$$

Note that v is computable because $\mathbb{K}^u$ is listable. Since u is a proper universal function, there is a computable unary map s such that

$$v_p = u_{s(p)}$$

for every $p \in \mathbb{N}$. Hence, for every $p \in \mathbb{N}$

- $s(p)$ is a u-index of function $\lambda\, x.x$ whenever $p \in \mathbb{K}^u$;
- $s(p)$ is a u-index of the undefined function whenever $p \notin \mathbb{K}^u$;

i.e.,

- if $p \in \mathbb{K}^u$ then range $\phi_{s(p)}^u = \mathbb{N}$ and so $k \in \text{range } \phi_{s(p)}^u$;
- if $p \notin \mathbb{K}^u$ then range $\phi_{s(p)}^u = \emptyset$ and so $k \notin \text{range } \phi_{s(p)}^u$.

That is,

$$\begin{aligned} p \in \mathbb{K}^u \quad &\text{iff} \quad k \in \text{range } \phi_{s(p)}^u \\ &\text{iff} \quad p \in \{q \in \mathbb{N} : k \in \text{range } \phi_{s(q)}^u\} \end{aligned}$$

for every $p \in \mathbb{N}$. Observe that

$$\begin{aligned} q \in \{q \in \mathbb{N} : k \in \text{range } \phi_{s(q)}^u\} \quad &\text{iff} \quad k \in \text{range } \phi_{s(q)}^u \\ &\text{iff} \quad s(q) \in \{q \in \mathbb{N} : k \in \text{range } \phi_q^u\} \end{aligned}.$$

So the decidability of $\{q \in \mathbb{N} : k \in \text{range } \phi_q^u\}$ implies the decidability of $\{q \in \mathbb{N} : k \in \text{range } \phi_{s(q)}^u\}$. Thus, the decidability of $\{q \in \mathbb{N} : k \in \text{range } \phi_q^u\}$ would entail the decidability of $\mathbb{K}^u$ (see Proposition 2.14). Since $\mathbb{K}^u$ is non-decidable, by Proposition 3.6, the given problem is non-decidable. ∇

Problem 4.3. Let u be a proper universal function. Show that the set

$$\{(p,q) \in \mathbb{N}^2 : \phi_p^u = \phi_q^u\}$$

is non-decidable.

Solution Observe that Rice's Theorem cannot be applied directly because the given set of indices is binary. We show that the problem

$$\text{Given } q \in \mathbb{N}, \text{ is } \lambda x.0 = \phi_q^u?$$

can be reduced to the problem

$$\text{Given } p,q \in \mathbb{N}, \text{ is } \phi_p^u = \phi_q^u?$$

Indeed, consider the map

$$s = \lambda q.(k,q) : \mathbb{N} \to \mathbb{N}^2$$

where k is an u-index of $\lambda x.0$. Then, it is immediate to see that s is computable and

$$q \in \{q \in \mathbb{N} : \lambda x.0 = \phi_q^u\} \quad \text{iff} \quad s(q) \in \{(p,q) \in \mathbb{N}^2 : \phi_p^u = \phi_q^u\}.$$

Since the first problem is non-decidable by Problem 4.1, then by Proposition 2.16, the second problem is also non-decidable. ∇

Problem 4.4. Let u be a proper universal function. Show that the set

$$\{p \in \mathbb{N} : p \in \text{dom } \phi_p^u\}$$

is non-decidable.

Solution In this case, we are unable to find a set $A \subseteq \mathscr{C}_1$ such that

$$\text{ind}^u(A) = \{p \in \mathbb{N} : p \in \text{dom } \phi_p^u\}$$

in order to apply Rice's Theorem. However, we are still able to prove this result using the reduction technique. Indeed, problem

$$\text{Given } p \in \mathbb{N}, \text{ does } p \in \mathbb{K}^u?$$

can be reduced to problem

$$\text{Given } p \in \mathbb{N}, \text{ does } p \in \{p \in \mathbb{N} : p \in \text{dom } \phi_p^u\}?$$

Observe that,

$$p \in \mathbb{K}^u \qquad \text{iff} \qquad p \in \operatorname{dom} \operatorname{diag}^u$$
$$\text{iff} \qquad p \in \operatorname{dom} \phi_p^u$$
$$\text{iff} \qquad p \in \{p \in \mathbb{N} : p \in \operatorname{dom} \phi_p^u\}.$$

for every $p \in \mathbb{N}$. Hence, if $\{p \in \mathbb{N} : p \in \operatorname{dom} \phi_p^u\}$ were decidable then so would be $\mathbb{K}^u$ contradicting Proposition 3.6.
$\qquad\qquad\qquad\qquad\qquad\qquad\qquad\qquad\qquad\qquad\qquad\qquad\qquad\qquad \nabla$

Problem 4.5. Show that there exists a non-proper computable universal function.

Solution Consider the function:

$$u = \lambda\, p, x \cdot \begin{cases} \text{undefined} & \text{if } p \text{ is even and } x = 0 \\[2mm] \operatorname{univ}\left(\dfrac{p}{2}, x\right) & \text{if } p \text{ is even and } x \neq 0 \\[2mm] K\left(\dfrac{p-1}{2}\right) & \text{if } p \text{ is odd and } x = 0 \\[2mm] \operatorname{univ}\left(L\left(\dfrac{p-1}{2}\right), x\right) & \text{if } p \text{ is odd and } x \neq 0 \end{cases}.$$

This function is clearly computable. Moreover:
(1) u is a universal function. Let $f \in \mathscr{C}_1$. Then, there is $q \in \mathbb{N}$ such that

$$f = \operatorname{univ}_q$$

since univ is a universal function. We now show that there is $p \in \mathbb{N}$ such that $f = u_p$. Consider two cases:
(a) Assume that $f(0)$ is not defined. Take $p = 2q$. We now show that

$$f = u_p.$$

Indeed,

$$\begin{cases} u_p(0) \text{ is undefined since } p \text{ is even} \\[2mm] u_p(x) = \operatorname{univ}\left(\dfrac{p}{2}, x\right) = \operatorname{univ}(q, x) = f(x) \text{ for } x \neq 0 \text{ since } p \text{ is even.} \end{cases}$$

(b) Assume that $f(0)$ is defined. Take

$$p = 2J(f(0), q) + 1.$$

Then,

$$u_p(0) \;=\; K\!\left(\frac{p-1}{2}\right)$$

$$=\; K\!\left(\frac{2J(f(0),q)}{2}\right) = f(0).$$

and, for $x \neq 0$

$$u_p(x) \;=\; \mathrm{univ}\!\left(L\!\left(\frac{p-1}{2}\right),x\right)$$

$$=\; \mathrm{univ}\!\left(L\!\left(\frac{2J(f(0),q)}{2}\right),x\right)$$

$$=\; \mathrm{univ}(q,x)$$

$$=\; f(x).$$

Hence $f = u_p$.

(2) u is not a proper function. Observe that

$$\mathrm{ind}^u(\{f \in \mathscr{C}_1 : 0 \in \mathrm{dom}\, f\}) = \{p \in \mathbb{N} : 0 \in \mathrm{dom}\, \phi_p^u\}$$

since u is a universal function. Moreover, this set is decidable since it is the set of all odd numbers. Assume, by contradiction, that u is proper. Since

$$\{f \in \mathscr{C}_1 : 0 \in \mathrm{dom}\, f\} \neq \emptyset \quad \text{and} \quad \{f \in \mathscr{C}_1 : 0 \in \mathrm{dom}\, f\} \neq \mathscr{C}_1,$$

by Rice's Theorem (Proposition 4.4), $\mathrm{ind}^u(\{f \in \mathscr{C}_1 : 0 \in \mathrm{dom}\, f\})$ is not a decidable set. Therefore, we reached a contradiction.   ▽

Problem 4.6. Let u be a proper universal function. Show that $\{p \in \mathbb{N} : \mathrm{dom}\, \phi_p^u = \emptyset\}$ is non-listable.

Solution Let $A = \{f \in \mathscr{C}_1 : \mathrm{dom}\, f = \emptyset\}$. Then $A = \{\lambda x.\,\text{undefined}\}$. Moreover,

$$\{p \in \mathbb{N} : \mathrm{dom}\, \phi_p^u = \emptyset\} = \mathrm{ind}^u(A)$$

since u is a universal function. Observe that

$$\lambda x.0 \notin A \quad \text{and} \quad \lambda x.\,\text{undefined} \subseteq \lambda x.0.$$

So, by Rice-Shapiro Theorem (Proposition 4.5) we conclude that $\mathrm{ind}^u(A) \notin \mathscr{L}_1$.   ▽

Problem 4.7. Let $f \in \mathscr{C}_1$ be such that $\mathrm{dom}\, f$ is either the empty set or an infinite set. Show that the problem

$$\text{Given } p \in \mathbb{N}, \text{ does } p \in \mathbb{K}^c?$$

is reducible to the problem

$$\text{Given } p \in \mathbb{N}, \text{ does } p \in \mathrm{ind}(f)?$$

Solution We must show that there is a map $s \in m\mathscr{C}_1$ such that

$$p \in \mathbb{K}^c \quad \text{iff} \quad s(p) \in \text{ind}(f).$$

for every $p \in \mathbb{N}$. We consider two cases:

(1) $f = \lambda x.\text{undefined}$. Let $g \in \mathscr{C}_1$ be such that $g \neq \lambda x.\text{undefined}$. Consider the function:

$$v = \lambda p, x . \begin{cases} g(x) & \text{if } p \in \mathbb{K} \\ \text{undefined} & \text{otherwise} \end{cases}.$$

Observe that

$$v = (\lambda x, y . x \times y) \circ \langle \chi_{\mathbb{K}}^{\text{P}} \circ \text{proj}_{[1]}^{\mathbb{N}^2}, g \circ \text{proj}_{[2]}^{\mathbb{N}^2} \rangle$$

which is clearly computable. Hence, by the *s-m-n* property for univ, there is $s \in m\mathscr{C}_1$ such that, for every $p \in \mathbb{N}$,

$$v_p = \phi_{s(p)}$$

So, for every $p \in \mathbb{N}$,

$$\begin{cases} \phi_{s(p)} = \lambda x . g(x) = g & \text{if } p \in \mathbb{K} \\ \phi_{s(p)} = \lambda x . \text{undefined} = f & \text{otherwise} \end{cases}.$$

Therefore,

$$p \in \mathbb{K}^c \quad \text{iff} \quad s(p) \in \text{ind}(f).$$

(2) dom f is an infinite set. Let v be the function computed by the following program:

```
function (p,x) (
    if stceval(P_{χ_{𝕂}^{P}}, p, x, −1) == 1 then
        while true do null
    else
        return P_f(x)
)
```

where P_f and $\text{P}_{\chi_{\mathbb{K}}^{\text{P}}}$ are programs that compute f and $\chi_{\mathbb{K}}^{\text{P}}$, respectively. Then, v is a computable function such that

$$v = \lambda p, x . \begin{cases} \text{undefined} & \text{if the execution of } \text{P}_{\chi_{\mathbb{K}}^{\text{P}}} \text{ on } p \text{ returns 1 in } x \text{ units of time} \\ f(x) & \text{otherwise} \end{cases}.$$

Thus, by the *s-m-n* property of univ, there is $s \in m\mathscr{C}_1$ such that, for every $p \in \mathbb{N}$,

$$v_p = \phi_{s(p)}.$$

If $p \in \mathbb{K}^c$ then, for every x, the execution of $P_{\chi_{\mathbb{K}}^p}$ on p does not return 1 in x units of time and so

$$\phi_{s(p)} = v_p = \lambda x . f(x) = f,$$

that is, $s(p) \in \mathrm{ind}(f)$. If $p \in \mathbb{K}$ then the execution of $P_{\chi_{\mathbb{K}}^p}$ on p returns 1 within t_p units of time for some $t_p \in \mathbb{N}$. Hence:

$$\phi_{s(p)} = v_p = \lambda x . \begin{cases} \text{undefined} & \text{if } x \geq t_p \\ f(x) & \text{if } x < t_p \end{cases}$$

which is clearly different from f because $\mathrm{dom}\, f$ is an infinite set. Thus, $s(p) \notin \mathrm{ind}(f)$.
∇

Problem 4.8. Consider the sequence of unary functions $\lambda\, k . g_k$ such that

$$g_k = \lambda x . \begin{cases} 1 & \text{if } x = 2k \\ \text{undefined} & \text{otherwise} \end{cases}$$

for every $k \in \mathbb{N}$ and the unary function g_∞ such that $\mathrm{graph}\, g_\infty = \bigcup_{k \in \mathbb{N}} \mathrm{graph}\, g_k$.

(1) Show that there is $s \in m\mathscr{C}_1$ such that $\theta_{s(k)} = g_k$ for each $k \in \mathbb{N}$.

(2) Analize the decidability and the listability of the set of indices of each of the following sets of computable unary functions:

- $A = \{f \in \mathscr{C}_1 : g_k \subseteq f \text{ for some } k \in \mathbb{N}\}$;
- $B = \{f \in \mathscr{C}_1 : g_k \nsubseteq f \text{ for every } k \in \mathbb{N}\}$;
- $C = \{f \in \mathscr{C}_1 : g_\infty \subseteq f\}$.

Solution (1) Recall how θ_p is obtained from any given $p \in \mathbb{N}$, as depicted in Figure 4.1. Let $\tau : f\mathscr{C}_1 \to \mathbb{N}$ be the inverse of the injective map $\lambda\, p . \theta_p : \mathbb{N} \to f\mathscr{C}_1$. The domain of each g_k is the singleton set $\{2k\}$. Thus, for each $k \in \mathbb{N}$, we have

$$\tau(g_k) = J(0, J(2k, 1)) + 1$$

because $\lambda z . b(1, z) = \lambda z . z$ and in this case $n = 1$, $m = 0$, $d = d_1 = x_1 = 2k$ and $r = y_1 = 1$. Therefore, the envisaged map is

$$s = \lambda\, k . J(0, J(2k, 1)) + 1.$$

This map is computable because it is a composition of computable maps, namely $\lambda z . z + 1, J, \lambda z . 0, \lambda z . 2z$ and $\lambda z . 1$.

(2) Decidability. The sets $\mathrm{ind}(A)$, $\mathrm{ind}(B)$ and $\mathrm{ind}(C)$ are all non-decidable, due to Rice's Theorem (Proposition 4.4):

- $A \neq \emptyset$ because, for instance, $g_0 \in A$;

- $A \neq \mathscr{C}_1$ because, for instance, $(\lambda x.\text{undefined}) \notin A$;

- $B = A^c$ and, so, $\emptyset \subsetneq B \subsetneq \mathscr{C}_1$ like A;

- $C \neq \emptyset$ because, for instance, $g_\infty \in C$ (note that g_∞ is the characteristic function of the set of even numbers);

- $C \neq \mathscr{C}_1$ because, for instance, $(\lambda x.\text{undefined}) \notin C$.

Listability. Set $\text{ind}(A)$ is listable thanks to Rice-Shapiro-McNaughton-Myhill Theorem (Proposition 4.8). Recall that this theorem states that, given $Z \subseteq \mathscr{C}_1$, the set $\text{ind}(Z)$ is listable iff there is a listable unary set P such that

$$Z = \bigcup_{p \in P} \{f \in \mathscr{C}_1 : \theta_p \subseteq f\} = \{f \in \mathscr{C}_1 : \theta_p \subseteq f \text{ for some } p \in P\}.$$

Clearly, the set

$$P = \{s(k) : k \in \mathbb{N}\}$$

is listable because it is the range of a computable function. Furthermore,

$$A = \{f \in \mathscr{C}_1 : \theta_{s(k)} \subseteq f \text{ for some } k \in \mathbb{N}\} = \{f \in \mathscr{C}_1 : \theta_p \subseteq f \text{ for some } p \in P\}$$

and, so, $\text{ind}(A)$ is listable.

Set $\text{ind}(B) = \text{ind}(A)^c$ is non-listable because otherwise, by Post's Theorem (Proposition 2.6), both $\text{ind}(A)$ and $\text{ind}(B)$ would be decidable and that is not true as we saw before.

Set $\text{ind}(C)$ is also non-listable because C does not fulfill the Rice-Shapiro condition since it is not empty and it does not contain any finite function. Therefore, thanks to the Rice-Shapiro Theorem (Proposition 4.5) $\text{ind}(C)$ is not be listable. ∇

Problem 4.9. Let $\lambda k. f_k : \mathbb{N} \to \mathscr{C}_1$ be a computable sequence. Show that there exist $r \in \text{m}\mathscr{C}_1$ such that:

1. $\text{range}\, r \subseteq \text{ind}(\{f_k : k \in \mathbb{N}\})$;

2. $(\text{range}\, r) \cap \text{ind}(f_k)$ is an infinite set for every $k \in \mathbb{N}$.

Solution Since $\lambda k. f_k$ is a computable sequence there exists $s \in \text{m}\mathscr{C}_1$ such that

$$f_k = \phi_{s(k)}$$

for every $k \in \mathbb{N}$. On the other hand, by Exercise 4.31, there exists $c \in \text{m}\mathscr{C}_2$ such that c_p is an injective map and

$$\text{range}(c_p) \subseteq \text{ind}(\phi_p)$$

for every $p \in \mathbb{N}$. That is,

$$\phi_{c_p(x)} = \phi_p$$

for every $p, x \in \mathbb{N}$. Let

$$r = c \circ \langle s \circ K, L \rangle.$$

Then, $r \in m\mathscr{C}_1$. Observe that:

(1) range $r \subseteq \text{ind}(\{f_k : k \in \mathbb{N}\})$. Indeed, assume that $p \in \text{range}\, r$. Then, there is ℓ such that $r(\ell) = p$. Thus,

$$
\begin{aligned}
\phi_p &= \phi_{r(\ell)} \\
&= \phi_{c(s(K(\ell)), L(\ell))} \\
&= \phi_{c_{s(K(\ell))}(L(\ell))} \\
&= \phi_{s(K(\ell))} \\
&= f_{K(\ell)}
\end{aligned}
$$

and so p is an index of $f_{K(\ell)}$. That is, $p \in \text{ind}(\{f_k : k \in \mathbb{N}\})$.

(2) $(\text{range}\, r) \cap \text{ind}(f_k)$ is an infinite set for every $k \in \mathbb{N}$. Let $k \in \mathbb{N}$. Recall that

$$\phi_{s(k)} = f_k$$

and

$$\text{range}\,(c_{s(k)}) \subseteq \text{ind}(\phi_{s(k)})$$

by hypoyhesis. Thus,

$$\text{range}\,(c_{s(k)}) \subseteq \text{ind}(f_k). \qquad (\dagger)$$

On the other hand,

$$\text{range}(c_{s(k)}) \text{ is an infinite set} \qquad (*)$$

since $c_{s(k)}$ is an injective map. Observe also that

$$\text{range}(c_{s(k)}) \subseteq \text{range}(r). \qquad (\ddagger)$$

Indeed, let $p \in \text{range}(c_{s(k)})$. Then, there is ℓ tal que $p = c_{s(k)}(\ell)$. Hence,

$$
\begin{aligned}
p &= c_{s(k)}(\ell) \\
&= c(s(k), \ell) \\
&= c(s(K(J(k, \ell))), L(J(k, \ell))) \\
&= r(J(k, \ell))
\end{aligned}
$$

and so $p \in \text{range}\,(r)$.

Therefore, by $(\dagger)$ e $(\ddagger)$, we can conclude that

$$\text{range}(c_{s(k)}) \subseteq \text{range}(r) \cap \text{ind}(f_k).$$

Furthermore,
$$\text{range}(r) \cap \text{ind}(f_k) \text{ é infinito.}$$

since $\text{range}(c_{s(k)})$ is an infinite set by $(*)$. ∇

Problem 4.10. Let U be a proper universal set. Show that, for any set $C \in \mathcal{L}_1$ there is a computable injective map $r : \mathbb{N} \rightarrow \mathbb{N}$ such that $\text{range}\, r \subseteq \text{ind}^U(C)$.

Solution It is convenient to consider two cases: the case $C \neq \emptyset$ and the case $C = \emptyset$. Here, we only consider the first, leaving the other to the reader. So, assume that $C \neq \emptyset$. Then, let
$$V = \{(p,x) : p \in \mathbb{K} \text{ and } x \in C\}.$$

Since $\mathbb{K}$ and C are both listable then V is listable because $V = \mathbb{K} \times C$ and the class of listable sets is closed for Cartesian product (Proposition 2.4). For each $p \in \mathbb{N}$,

- if $p \in \mathbb{K}$ then $V_p = C$;

- if $p \notin \mathbb{K}$ then $V_p = \emptyset$.

As U is a proper universal set then there is a computable map $s : \mathbb{N} \rightarrow \mathbb{N}$ such that
$$V_p = U_{s(p)}$$

for every $p \in \mathbb{N}$. Then

- if $p \in \mathbb{K}$ then $\Gamma^U_{s(p)} = V_p = C$;

- if $p \notin \mathbb{K}$ then $\Gamma^U_{s(p)} = V_p = \emptyset$.

Given an enumeration h of $\mathbb{K}$, then
$$\text{range}(s \circ h) \subseteq \text{ind}^U(C).$$

In fact, if $p \in \text{range}(s \circ h)$ then there is $p' \in \text{range}\, h$ such that $s(p') = p$. But, as h is an enumeration of $\mathbb{K}$, then $p' \in \mathbb{K}$ which means that
$$\Gamma^U_p = \Gamma^U_{s(p')} = V_{p'} = C.$$

So, $p \in \text{ind}^U(C)$. Furthermore, $\text{range}(s \circ h)$ is infinite as we show now. Indeed, observe that

- if $p \in \mathbb{K}$ then, as h is an enumeration of $\mathbb{K}$,
$$s(p) \in s(\mathbb{K}) = s(\text{range}\, h) = \text{range}(s \circ h);$$

- if $p \notin \mathbb{K}$ then $\Gamma^U_{s(p)} = \emptyset$ and, as $C \neq \emptyset$, then $s(p) \notin \text{ind}^U(C)$, and, so,

$$s(p) \notin \text{range}(s \circ h)$$

because $\text{range}(s \circ h) \subseteq \text{ind}^U(C)$.

Hence, we have that

$$p \in \mathbb{K} \quad \text{iff} \quad s(p) \in \text{range}(s \circ h).$$

Therefore, the problem

$$\text{Given } p \in \mathbb{N}, \text{ does } p \in \mathbb{K}?$$

is reducible to the problem

$$\text{Given } q \in \mathbb{N}, \text{ does } q \in \text{range}(s \circ h)?$$

Thus, if $\text{range}(s \circ h)$ is a finite set then it is decidable and, so, by Proposition 2.14, $\mathbb{K}$ is also decidable, in contradiction with Proposition 3.6. Thus, $\text{range}(s \circ h)$ is infinite.

So, we were able to find computable maps s and h such that:

- $\text{range}(s \circ h) \subseteq \text{ind}^U(C)$;

- $\text{range}(s \circ h)$ is infinite.

Thus, $\text{range}(s \circ h)$ is an infinite listable set. Hence, using Proposition 2.9, there is an injective enumeration r of $\text{range}(s \circ h)$. Then, r fulfills the requirements of the thesis.
∇

Problem 4.11. Let $u \in F_2$ be a function and let u' be the function:

$$u' = \lambda \, p, x \cdot \begin{cases} u\left(\dfrac{p}{2}, x\right) & \text{if } p \text{ is even} \\[2mm] u\left(\dfrac{p-1}{2}, x\right) & \text{if } p \text{ is odd.} \end{cases}$$

Show that there a morphism from u to u'. Moreover, show that u' is a proper universal function whenever so is u.

Solution (1) Consider the map

$$s' = \lambda \, x \cdot \begin{cases} 2x & \text{if } x \text{ is even} \\ 2x+1 & \text{if } x \text{ is odd.} \end{cases} : \mathbb{N} \to \mathbb{N}$$

We start by observing that s' is a computable map. Moreover, for every $p \in \mathbb{N}$,

$$u'_{s'(p)} = u_p.$$

If p is even, then $u'_{s'(p)} = u'_{2p} = u_p$. If p is odd, then $u'_{s'(p)} = u'_{2p+1} = u_p$. Therefore,

$$s' : u \to u'$$

is a morphism from u to u'.

(2) u' is a proper universal function. Let $v' \in \mathscr{C}_2$ be any function. Since u is a proper universal function, by Proposition 4.10, there is morphism $s : v' \to u$. Thus, $s' \circ s : v' \to u'$ is a morphism and so, by the same proposition, u' is a proper universal function. ∇

Problem 4.12. Let u be a proper universal function and $v \in \mathscr{C}_2$. Show that there is $p \in \mathbb{N}$ such that $v_p = u_p$.

Solution As u is a proper universal function there $s \in m\mathscr{C}_1$ such that

$$v_k = u_{s(k)}$$

for every $k \in \mathbb{N}$. By the Recursion Theorem (Proposition 4.17) there is $p \in \mathbb{N}$ such that

$$u_{s(p)} = u_p.$$

Hence, it follows that

$$u_p = u_{s(p)} = v_p.$$

∇

Problem 4.13. Prove Rice's Theorem using the Recursion Theorem.

Solution Let u be a proper universal function and $A \subseteq \mathscr{C}_1$. We have to show that

$$\text{ind}^u(A) \text{ is decidable} \quad \text{iff} \quad \text{either } A = \emptyset \text{ or } A = \mathscr{C}_1.$$

($\to$) Assume, by contradiction, that $A \neq \emptyset$ and $A \neq \mathscr{C}_1$. Let $a \in \text{ind}^u(A)$ and $b \notin \text{ind}^u(A)$. Consider the map

$$f = \lambda x. \begin{cases} a & \text{if } x \notin \text{ind}^u(A) \\ b & \text{otherwise} \end{cases} : \mathbb{N} \to \mathbb{N}.$$

Then, it follows that

$$x \in \text{ind}^u(A) \quad \text{iff} \quad f(x) \notin \text{ind}^u(A)$$

for every $x \in \mathbb{N}$. Observe that f is computable, since $\text{ind}^u(A)$ is decidable. So, by the Recursion Theorem (Proposition 4.17), there is $p \in \mathbb{N}$ such that

$$\phi^u_{f(p)} = \phi^u_p.$$

Therefore,

$$
\begin{aligned}
\phi_p^u \in A \quad &\text{iff} \quad p \in \text{ind}^u(A) \\
&\text{iff} \quad f(p) \notin \text{ind}^u(A) \\
&\text{iff} \quad \phi_{f(p)}^u \notin A
\end{aligned}
$$

which is a contradiction. Hence, either $A = \emptyset$ or $A = \mathscr{C}_1$.

($\leftarrow$) It is enough to observe that $\text{ind}^u(\emptyset) = \emptyset$ and $\text{ind}^u(\mathscr{C}_1) = \mathbb{N}$. 　　　　$\triangledown$

Problem 4.14. Show that there is $p \in \mathbb{N}$ such that

$$
p + 1 \approx^{\text{univ}} p + 2.
$$

Solution Consider the computable map

$$
s = \lambda x . x + 1.
$$

Then, by Proposition 4.18, there is $q \in \mathbb{N}^+$ such that

$$
\phi_q = \phi_{s(q)}.
$$

That is,

$$
q \approx^{\text{univ}} q + 1.
$$

Let $p = q - 1$. Then

$$
p + 1 = q \approx^{\text{univ}} q + 1 = p - 2.
$$

　　　　$\triangledown$

Supplementary Exercises

Exercise 4.25. Let u be a proper universal function. Provide $A \subseteq \mathscr{C}_1$ such that A fulfills the Rice-Shapiro condition but $\text{ind}^u(A)$ is non-listable.

Exercise 4.26. Let u be a proper universal function. Show that $\{p \in \mathbb{N} : \phi_p^u = \lambda x . x + 1\}$ is not a listable set.

Exercise 4.27. Show that $\{p \in \mathbb{N} : 0 \in \text{dom}\, \phi_p\}$ is a listable set.

Exercise 4.28. Let u be a proper universal function. Verify which of the following sets are listable or not:

1. $\{p \in \mathbb{N} : \phi_p^u \text{ is total}\}$;

2. $\{p \in \mathbb{N} : \phi_p^u \text{ is not injective}\}$;

3. $\{p \in \mathbb{N} : \phi_p^u \text{ is injective}\}$;

4. $\{p \in \mathbb{N} : \operatorname{dom} \phi_p = \operatorname{range} \phi_p\}$.

Exercise 4.29. Let u be a proper universal function and $\operatorname{ind}^u(A) \in \mathcal{L}_1$. Show that:

1. If $A \neq \emptyset$ then there is a finite function in A.

2. If the undefined function is in A then $A = \mathscr{C}_1$.

3. Any extension of a function in A is also in A.

Exercise 4.30. Let u be a proper universal function. A set $C \subseteq \mathbb{N}$ is said to be u-*extensional* if, for every $p, q \in \mathbb{N}$, $p \in C$ if and only if $q \in C$ whenever $\phi_p^u = \phi_q^u$. Show that:

1. $\{p \in \mathbb{N} : k \in \operatorname{dom} \phi_p^u\}$ is u-extensional, for any given $k \in \mathbb{N}$.

2. $\{p \in \mathbb{N} : p \in \operatorname{dom} \phi_p^u\}$ is not u-extensional.

3. C is u-extensional if and only if there is $A \subseteq \mathscr{C}_1$ such that $C = \operatorname{ind}^u(A)$.

4. A u-extensional set C is decidable if and only if $C = \emptyset$ or is $C = \mathbb{N}$.

Exercise 4.31. Let u be a proper universal function. Then, there is $c \in m\mathscr{C}_2$ such that, for each $p \in \mathbb{N}$, the following conditions hold:

- c_p is injective;

- $\operatorname{range} c_p \subseteq \operatorname{ind}^u(\phi_p^u)$.

Exercise 4.32. Let u be a proper universal function. Show that there is $p \in \mathbb{N}$ such that:

1. $\phi_p^u = \lambda x. \begin{cases} p & \text{if } x = p \\ \text{undefined} & \text{otherwise} \end{cases}$;

2. $\phi_p^u = \lambda x. p^2$.

Exercise 4.33. Show that, in general, it is not the case that for all $s, r \in m\mathscr{C}_1$ exists p such that
$$s(p) \approx^{\text{univ}} r(p).$$

Exercise 4.34. Let U be a proper universal set, $s \in m\mathscr{C}_1$ and $q \in \mathbb{N}$. Show that there is $p \geq q$ such that
$$\Gamma_{s(p)}^U = \Gamma_p^U.$$

Exercise 4.35. Let u be a proper universal function and $e : \mathbb{N} \to \mathbb{N}$ such that

1. if $i < j$ then $e(i) < e(j)$;

2. if $i \neq j$ then $\phi_{e(i)}^u \neq \phi_{e(j)}^u$;

3. $e(i) = \min(\operatorname{ind}^u(\phi_{e(i)}^u))$.

Show that e is not a computable map.

Chapter 5

Reducibility

It is now time to develop the underlying theory of reducibility on sets (instead of on problems as presented in Section 2.3). There are several, not necessarily equivalent, ways to make precise the notion of reduction. In this chapter, we introduce the notion of reducibility based on a computable map (many-to-one reducibility). We also discuss many-to-one complete sets as well as effectively non-listable sets.

5.1 Many-to-One Reducibility

In this section we introduce the basic concept and study its properties.

Definition 5.1. Let $C_1, C_2 \in \mathscr{S}_1$. Then, C_1 is said to be *many-to-one reducible* or m-*reducible* to C_2, written

$$C_1 \leq_m C_2,$$

if there exists $s \in m\mathscr{C}_1$ such that

$$x \in C_1 \text{ iff } s(x) \in C_2$$

for every $x \in \mathbb{N}$.

Exercise 5.1. Show that $\leq_m$ is a preorder (that is, a reflexive and transitive relation) over $\mathscr{S}_1$.

Remark 5.1. Given $C_1, C_2 \in \mathscr{S}_1$ and $s \in m\mathscr{C}_1$, we write

$$s : C_1 \leq_m C_2$$

whenever $x \in C_1$ if and only if $s(x) \in C_2$ for every $x \in \mathbb{N}$.

Exercise 5.2. Let $s : C_1 \leq_m C_2$. Verify that:

1. $s(C_1) \subseteq C_2$;

2. $C_1 \subseteq s^{-1}(C_2)$;

3. $\chi_{C_1} = \chi_{C_2} \circ s$;

4. $\chi_{C_1}^p = \chi_{C_2}^p \circ s$.

Proposition 5.1. Let $C, D \in \mathscr{S}_1$. Then:

1. $C \leq_m D$ iff $C^c \leq_m D^c$;

2. if $C \leq_m D$ and D is decidable then C is decidable;

3. if $C \leq_m D$ and D is listable then C is listable;

4. if C is decidable and $\emptyset \subsetneq D \subsetneq \mathbb{N}$ then $C \leq_m D$;

5. $C \leq_m \mathbb{N}$ iff $C = \mathbb{N}$;

6. $C \leq_m \emptyset$ iff $C = \emptyset$;

7. $\mathbb{N} \leq_m D$ iff $D \neq \emptyset$;

8. $\emptyset \leq_m D$ iff $D \neq \mathbb{N}$.

Proof. (1) For the proof of the left to right direction assume that $s : C \leq_m D$. Then, for every $x \in \mathbb{N}$,
$$x \in C \quad \text{iff} \quad s(x) \in D,$$
hence
$$x \notin C \quad \text{iff} \quad s(x) \notin D$$
and so
$$x \in C^c \quad \text{iff} \quad s(x) \in D^c.$$

Therefore, $s : C^c \leq_m D^c$. For the proof of the right to left direction assume that $s : C^c \leq_m D^c$. Then,
$$s : (C^c)^c \leq_m (D^c)^c$$
and so $s : C \leq_m D$.

(2) Assume that D is decidable and let $s : C \leq_m D$. Then, χ_D is computable. Thus, $\chi_C = \chi_D \circ s$, by Exercise 5.2. Therefore, χ_C is computable and, so, C is decidable.

(3) Assume that D is listable and let $s : C \leq_m D$. Then, χ_D^p is computable. Thus, $\chi_C^p = \chi_D^p \circ s$, by Exercise 5.2. Hence, χ_C^p is computable and, so, C is listable.

(4) Let $\emptyset \subsetneq D \subsetneq \mathbb{N}$ with $d \in D$ and $e \in D^c$. Then, since C is decidable by hypothesis, the map

$$s = \lambda x. \begin{cases} d & \text{if } x \in C \\ e & \text{otherwise} \end{cases} : \mathbb{N} \to \mathbb{N}$$

is computable. Furthermore,

$$x \in C \quad \text{iff} \quad s(x) \in D$$

for every $x \in \mathbb{N}$. Hence $s : C \leq_m D$.

(5) For the proof of the left to right direction assume that $s : C \leq_m \mathbb{N}$. Then:

$$x \in C \quad \text{iff} \quad s(x) \in \mathbb{N} \quad \text{iff} \quad \text{true}$$

for every $x \in \mathbb{N}$. Therefore $C = \mathbb{N}$. The proof of the right to left direction follows immediately thanks to the reflexivity of $\leq_m$.

(6) It is enough to observe that

$$C \leq_m \emptyset \quad \text{iff} \quad C^c \leq_m \emptyset^c \quad \text{iff} \quad C^c \leq_m \mathbb{N} \quad \text{iff} \quad C^c = \mathbb{N} \quad \text{iff} \quad C = \emptyset.$$

(7) For the proof of the left to right direction assume that $s : \mathbb{N} \leq_m D$. Then,

$$s(x) \in D \quad \text{iff} \quad x \in \mathbb{N} \quad \text{iff} \quad \text{true}$$

for every $x \in \mathbb{N}$. Thus, $s(\mathbb{N}) \subseteq D$ and, so, since s is a map, $D \neq \emptyset$. For the proof of the right to left direction assume that $D \neq \emptyset$. Let $d \in D$. Consider the computable map

$$s = \lambda x.d : \mathbb{N} \to \mathbb{N}.$$

Clearly,

$$x \in \mathbb{N} \quad \text{iff} \quad s(x) \in D$$

for every $x \in \mathbb{N}$. Hence, $s : \mathbb{N} \leq_m D$.

(8) It is enough to observe that

$$\emptyset \leq_m D \quad \text{iff} \quad \emptyset^c \leq_m D^c \quad \text{iff} \quad \mathbb{N} \leq_m D^c \quad \text{iff} \quad D^c \neq \emptyset \quad \text{iff} \quad D \neq \mathbb{N}.$$

$\square$

The following result is not surprising, since the complement of a listable non-decidable set should not be many-to-one reducible to it.

Proposition 5.2. Let C be a listable non-decidable set. Then:

1. $C^c \nleq_m C$;

2. $C \not\leq_m C^c$.

Proof. (1) Given that C is a listable set, if $C^c \leq_m C$ then, by Proposition 5.1, C^c would also be a listable set and so C would be decidable by Post's Theorem (Proposition 2.6) contradicting the undecidability of C.

(2) Again by contradiction, assume that $C \leq_m C^c$. Then, by Proposition 5.1, $C^c \leq_m (C^c)^c$ and, therefore, $C^c \leq_m C$, contradicting statement (1).   □

The following result confirms that m-reducibility is useful for comparing the difficulty in semideciding problems. It entails that no semidecidable problem is harder to solve than the problem

$$\text{Given } p \in \mathbb{N}, \text{ does } p \in \mathbb{K}^u?$$

Proposition 5.3. Let u be a proper universal function and $C \in \mathscr{S}_1$. Then,

$$C \text{ is listable} \quad \text{iff} \quad C \leq_m \mathbb{K}^u.$$

Proof. ($\leftarrow$) Assume that $C \leq_m \mathbb{K}^u$. Then, C is a listable set, thanks to Proposition 5.1, since $\mathbb{K}^u$ is listable.

($\rightarrow$) Assume that C is a listable set. Then, the binary function

$$v = \lambda \, p, x \, . \, \chi_C^p(p).$$

is computable. Furthermore,

$$(p,x) \in \mathrm{dom}\, v \quad \text{iff} \quad p \in C$$

for every $p, x \in \mathbb{N}$. Since u is a proper universal function, there is $s \in m\mathscr{C}_1$ such that

$$v(p,x) = u(s(p),x)$$

for every $p, x \in \mathbb{N}$. Hence,

$$
\begin{array}{llll}
p \in C & \text{iff} & (p, s(p)) \in \mathrm{dom}\, v \\
        & \text{iff} & (s(p), s(p)) \in \mathrm{dom}\, u \\
        & \text{iff} & s(p) \in \mathrm{dom\,diag}^u \\
        & \text{iff} & s(p) \in \mathbb{K}^u
\end{array}
$$

and, so, $s : C \leq_m \mathbb{K}^u$.   □

5.2 Many-to-One Degrees

In this section we classify sets that can be many-to-one reducible to each other.

Definition 5.2. Given $C_1, C_2 \in \mathscr{S}_1$, we say that C_1 and C_2 are *many-to-one equivalent* or m-*equivalent*, written

$$C_1 \equiv_m C_2,$$

if $C_1 \leq_m C_2$ and $C_2 \leq_m C_1$.

Exercise 5.3. Show that $\equiv_m$ is an equivalence relation over $\mathscr{S}_1$.

Remark 5.2. Given $C \in \mathscr{S}_1$, we denote by

$$[C]_m$$

the equivalence class $\{D \in \mathscr{S}_1 : C \equiv_m D\}$ of C with respect to the equivalence relation $\equiv_m$.

In $[C]_m$ we find the sets that, according to m-reducibility, are as difficult to (semi) decide as C.

Exercise 5.4. Let u and u' be proper universal functions. Show that:

$$[\mathbb{K}^u]_m = [\mathbb{K}^{u'}]_m.$$

Definition 5.3. We denote by

$$\mathsf{D}_m$$

the quotient set $\mathscr{S}_1/\equiv_m$ of $\mathscr{S}_1$ by $\equiv_m$. The elements of D_m are called m-*degrees*. Moreover, we denote by

$$\mathsf{d}_m$$

the surjective map $\lambda C.[C]_m : \mathscr{S}_1 \to \mathscr{S}_1/\equiv_m$ that assigns to each unary set its m-degree.

The relation $\leq_m$ over $\mathscr{S}_1$ induces, in a natural way, a relation also named $\leq_m$ over D_m.

Definition 5.4. Let $\leq_m$ be the binary relation over D_m such that:

$$\mathsf{d}_m(C) \leq_m \mathsf{d}_m(D) \quad \text{if} \quad C \leq_m D.$$

Exercise 5.5. Verify that the relation $\leq_m$ over D_m is well defined and is a partial order.

Exercise 5.6. Let **c** and **d** be m-degrees. Verify that:

1. $\mathbf{c} \leq_m \mathbf{d}$ iff $C \leq_m D$ for some $C \in \mathbf{c}$ and $D \in \mathbf{d}$;

2. $\mathbf{c} \leq_m \mathbf{d}$ iff $C \leq_m D$ for every $C \in \mathbf{c}$ and $D \in \mathbf{d}$.

Remark 5.3. The following notation becomes handy when exploring the class of m-degrees:

- $\mathbf{o} = d_m(\emptyset)$;

- $\mathbf{n} = d_m(\mathbb{N})$;

- $\mathbf{0}_m = d_m(\{0\})$;

- $\mathbf{0}'_m = d_m(\mathbb{K})$.

Definition 5.5. An m-degree is said to be *decidable* if it contains a decidable set, and it is said to be *listable* if it contains a listable set.

Exercise 5.7. Show that $\mathbf{o}$, $\mathbf{n}$ and $\mathbf{0}_m$ are decidable degrees and $\mathbf{0}'_m$ is a listable degree.

Remark 5.4. Denote by

$$D_m^{\mathscr{D}}$$

the set of decidable m-degrees and by

$$D_m^{\mathscr{L}}$$

the set of listable m-degrees.

Exercise 5.8. Show that every element of a decidable m-degree is a decidable set.

The following results collect in the language of m-degrees the relevant results established in the previous section and provides a useful insight on the partial order on m-degrees.

Proposition 5.4 (Structure of $D_m^{\mathscr{D}}$). The following assertions about m-degrees hold:

1. $\mathbf{o} = \{\emptyset\}$;

2. $\mathbf{n} = \{\mathbb{N}\}$;

3. $\mathbf{0}_m = \{C : \emptyset \subsetneq C \subsetneq \mathbb{N} \text{ and } C \in \mathscr{D}_1\}$;

4. $D_m^{\mathscr{D}} = \{\mathbf{o}, \mathbf{n}, \mathbf{0}_m\}$.

Proof. (1) and (2) are straightforward consequences of Proposition 5.1.

(3) For the left to right inclusion, let $C \in \mathbf{0}_m$. Since $\mathbf{0}_m = d_m(\{0\}) = [\{0\}]_m$ then $C \in [\{0\}]_m$. That is,

$$C \leq_m \{0\} \quad \text{and} \quad \{0\} \leq_m C.$$

Hence:

(a) $C \in \mathscr{D}_1$, by Proposition 5.1, since $\{0\}$ is a decidable set and $C \leq_m \{0\}$.

(b) $C \neq \emptyset$, by Proposition 5.1, since $\{0\} \leq_m C$ and $\{0\} \neq \emptyset$.

(c) $C \neq \mathbb{N}$, by Proposition 5.1, since $\{0\} \leq_m C$ and $\{0\} \neq \mathbb{N}$.

For the right to left inclusion, let $C \in \mathscr{D}_1$ be such that $C \neq \emptyset$ and $C \neq \mathbb{N}$. Then, by Proposition 5.1,

$$\{0\} \leq_m C \quad \text{and} \quad C \leq_m \{0\}.$$

Therefore, $C \in [\{0\}]_m = \mathbf{0}_m$.

(4) For the left to right inclusion, let $\mathbf{d} \in D_m^{\mathscr{D}}$. Hence, $\mathbf{d}$, by definition of $D_m^{\mathscr{D}}$, contains a decidable set D. Thus,

$$\mathbf{d} = [D]_m = d_m(D).$$

We consider three cases:

(a) $D = \emptyset$. Then, $\mathbf{d} = d_m(\emptyset) = \mathbf{o}$ and so $\mathbf{d} \in \{\mathbf{o}, \mathbf{n}, \mathbf{0}_m\}$.

(b) $D = \mathbb{N}$. Therefore, $\mathbf{d} = d_m(\mathbb{N}) = \mathbf{n}$ and so $\mathbf{d} \in \{\mathbf{o}, \mathbf{n}, \mathbf{0}_m\}$.

(c) $D \neq \emptyset$ and $D \neq \mathbb{N}$. Then, by (3), $D \in \mathbf{0}_m$ and so $\mathbf{d} = d_m(D) = [D]_m = \mathbf{0}_m$. Thus $\mathbf{d} \in \{\mathbf{o}, \mathbf{n}, \mathbf{0}_m\}$

For the right to left inclusion, it is enough to show that $\mathbf{o}$, $\mathbf{n}$ and $\mathbf{0}_m$ contain, each one, a decidable set. But, by definition, $\emptyset \in \mathbf{o}$, $\mathbb{N} \in \mathbf{n}$ and $\{0\} \in \mathbf{0}_m$. Furthermore, $\emptyset$, $\mathbb{N}$ and $\{0\}$ are decidable sets. Thus, $\{\mathbf{o}, \mathbf{n}, \mathbf{0}_m\} \subseteq D_m^{\mathscr{D}}$. $\square$

Contrarily to the precise characterization of $D_m^{\mathscr{D}}$ the situation for $D_m^{\mathscr{L}}$ is not so clear. Nevertheless, we can establish the following result that we do not prove since it follows straightforwardly by Proposition 5.1.

Proposition 5.5 (Structure of $D_m^{\mathscr{L}}$). Let $\mathbf{c}$ and $\mathbf{d}$ be m-degrees. Then:

1. $\mathbf{d} \subset \mathscr{L}_1$ if and only if $\mathbf{d} \in D_m^{\mathscr{L}}$;

2. if $\mathbf{c} \leq_m \mathbf{d}$ and $\mathbf{d} \in D_m^{\mathscr{L}}$ then $\mathbf{c} \in D_m^{\mathscr{L}}$;

3. $\mathbf{0}'_m \in D_m^{\mathscr{L}}$;

4. $\mathbf{c} \leq_m \mathbf{0}'_m$ if and only if $\mathbf{c} \in D_m^{\mathscr{L}}$.

Exercise 5.9. Show that $D_m^{\mathcal{Q}} \subsetneq D_m^{\mathcal{L}}$.

Additional statements about m-degrees are proposed in the following exercise:

Exercise 5.10. Let $\mathbf{c}, \mathbf{d}$ be m-degrees. Show that:

1. $\mathbf{d} \subset \mathcal{D}_1$ if and only if $\mathbf{d} \in D_m^{\mathcal{Q}}$;

2. if $\mathbf{c} \leq_m \mathbf{d}$ and $\mathbf{d} \in D_m^{\mathcal{Q}}$ then $\mathbf{c} \in D_m^{\mathcal{Q}}$;

3. $\mathbf{o} \leq_m \mathbf{d}$ whenever $\mathbf{d} \neq \mathbf{n}$;

4. $\mathbf{n} \leq_m \mathbf{d}$ whenever $\mathbf{d} \neq \mathbf{o}$;

5. $\mathbf{0}_m \leq_m \mathbf{d}$ whenever $\mathbf{d} \neq \mathbf{o}$ and $\mathbf{d} \neq \mathbf{n}$.

We now establish some results related to the cardinality of m-degrees.

Proposition 5.6. The following assertions about m-degrees hold:

1. Every $\mathbf{d} \in D_m$ is countable;

2. D_m is not countable;

3. $D_m^{\mathcal{L}}$ is countable.

Proof. (1) Let $\mathbf{d} \in D_m$. Then, there is $C \in \mathscr{S}_1$ such that $\mathbf{d} = [C]_m$. If $D \in \mathbf{d}$ then $D \equiv_m C$, that is, $D \leq_m C$ and $C \leq_m D$. Hence, there are $s_1, s_2 \in m\mathscr{C}_1$ such that $s_1 : D \leq_m C$ and $s_2 : C \leq_m D$. As $m\mathscr{C}_1$ is denumerable then $\mathbf{d}$ is either finite or denumerable, that is, $\mathbf{d}$ is countable. Note that the same map cannot be used to reduce two different sets to C. Indeed, if $s_1 : D \leq_m C$ and $s_1 : D' \leq_m C$ then

$$d \in D \quad \text{iff} \quad s_1(d) \in C \quad \text{iff} \quad d \in D'$$

and so $D = D'$.

(2) Observe that

$$\bigcup_{\mathbf{d} \in D_m} \mathbf{d} = \mathscr{S}_1 = \wp\mathbb{N}$$

which is not countable. We know that the countable union of countable sets is countable. Taking into account (1), each $\mathbf{d} \in D_m$ is countable. Therefore, the only way for $\bigcup_{\mathbf{d} \in D_m} \mathbf{d}$ to be not countable is that D_m is not countable.

(3) This is a direct consequence of $\mathcal{L}_1$ being countable. $\square$

In any given partial order (S, R), an *upper bound* for a pair $s_1, s_2 \in S$ is an element $s' \in S$ such that $s_1 R s'$ and $s_2 R s'$. An upper bound $s \in S$ of s_1, s_2, is said to be the *least upper bound* or the *supremum* of s_1, s_2 whenever every upper bound s' of s_1, s_2 is such that $s R s'$. A partial order is said to be an *upper semilattice* whenever there is a least upper bound for every pair of elements.

Proposition 5.7. The pair $(D_m, \leq_m)$ is an upper semilattice.

Proof. Observe that $\leq_m$ is a partial order over D_m by Exercise 5.5. Let $\mathbf{d}_1$ and $\mathbf{d}_2$ be any m-degrees. Choose $D_1 \in \mathbf{d}_1$ and $D_2 \in \mathbf{d}_2$. Take

$$\mathbf{d} = [D]_m$$

where

$$D = \{2x : x \in D_1\} \cup \{2x+1 : x \in D_2\}.$$

That is, D is a disjoint union of D_1 and D_2. We must verify that $\mathbf{d}$ is indeed a least upper bound of $\mathbf{d}_1$ and $\mathbf{d}_2$, That is, that the two following conditions hold:

1. $\mathbf{d}_1 \leq_m \mathbf{d}$ and $\mathbf{d}_2 \leq_m \mathbf{d}$;

2. if $\mathbf{d}_1 \leq_m \mathbf{c}$ and $\mathbf{d}_2 \leq_m \mathbf{c}$ then $\mathbf{d} \leq_m \mathbf{c}$, for every m-degree $\mathbf{c}$;

as we now show:

(1) Let $s_1 : \mathbb{N} \to \mathbb{N}$ and $s_2 : \mathbb{N} \to \mathbb{N}$ be computable maps such that $s_1 = \lambda x.2x$ and $s_1 = \lambda x.2x+1$. Then, for $i = 1, 2$,

$$x \in D_i \quad \text{iff} \quad s_i(x) \in D$$

for every $x \in \mathbb{N}$. Thus,

$$D_1 \leq_m D \text{ and } D_2 \leq_m D$$

and the thesis follows.

(2) Assume that $\mathbf{d}_1 \leq_m \mathbf{c}$ and $\mathbf{d}_2 \leq_m \mathbf{c}$. Let $D_1 \in \mathbf{d}_1$, $D_2 \in \mathbf{d}_2$ and $C \in \mathbf{c}$. Then, $D_1 \leq_m C$ and $D_2 \leq_m C$ by Exercise 5.6. Let $s_i : \mathbb{N} \to \mathbb{N}$ be computable maps such that

$$x \in D_i \quad \text{iff} \quad s_i(x) \in C, \text{ for every } x \in \mathbb{N}.$$

Define map

$$s = \lambda x. \begin{cases} s_1\left(\dfrac{x}{2}\right) & \text{if } x \text{ is even} \\ s_2\left(\dfrac{x-1}{2}\right) & \text{otherwise} \end{cases}.$$

It remains to prove that

$$x \in D \quad \text{iff} \quad s(x) \in C$$

for every $x \in \mathbb{N}$. Indeed:

($\to$) Assume that $x \in D$. Consider two cases:
(a) x is $2y$ for some $y \in \mathbb{N}$. Then, $y \in D_1$ and so $s_1(y) \in C$. Since $s_1(y) = s(x)$ then $s(x) \in C$.

(b) x is $2y+1$ for some $y \in \mathbb{N}$. Then, $y = (x-1)/2 \in D_2$ and so $s_2(y) = s_2((x-1)/2) = s(x) \in C$.

($\leftarrow$) Assume that $s(x) \in C$. We have two cases:

(a) x is even. Then, $s(x) = s_1(x/2)$ and so $x/2 \in D_1$. Hence, $x \in D$.

(a) x is odd. Then, $s(x) = s_2((x-1)/2)$ and so $(x-1)/2 \in D_2$. Hence, $x \in D$.

Therefore, $D \leq_m C$. $\square$

Remark 5.5. Given $\mathbf{c}, \mathbf{d} \in D_m$, we denote by

$$\mathbf{c} \sqcup_m \mathbf{d}$$

a least upper bound of $\mathbf{c}$ and $\mathbf{d}$.

In the following result, we analyse the listability and the decidability of $\mathbf{c} \sqcup_m \mathbf{d}$ when $\mathbf{c}$ and $\mathbf{d}$ are assumed to be listable and decidable, respectively.

Proposition 5.8. Let $\mathbf{c}$ and $\mathbf{d}$ be m-degrees. Then,

1. if $\mathbf{c}, \mathbf{d} \in D_m^{\mathscr{L}}$ then $\mathbf{c} \sqcup_m \mathbf{d} \in D_m^{\mathscr{L}}$;

2. if $\mathbf{c}, \mathbf{d} \in D_m^{\mathscr{D}}$ then $\mathbf{c} \sqcup_m \mathbf{d} \in D_m^{\mathscr{D}}$;

3. if $\mathbf{c} \leq_m \mathbf{d}$ then $\mathbf{c} \sqcup_m \mathbf{d} = \mathbf{d}$.

Proof. (1) Observe that

$$\mathbf{d} \in D_m^{\mathscr{L}} \quad \text{iff} \quad \mathbf{d} \leq_m \mathbf{0}_m'$$

by Proposition 5.3. Assume that $\mathbf{c}, \mathbf{d} \in D_m^{\mathscr{L}}$. Hence,

$$\mathbf{c} \leq_m \mathbf{0}_m' \text{ and } \mathbf{d} \leq_m \mathbf{0}_m'$$

and so, since $\mathbf{c} \sqcup_m \mathbf{d}$ is a least upper bound of $\mathbf{c}$ and $\mathbf{d}$ we can conclude that $\mathbf{c} \sqcup_m \mathbf{d} \leq_m \mathbf{0}_m'$. Therefore, $\mathbf{c} \sqcup_m \mathbf{d} \in D_m^{\mathscr{L}}$ by Proposition 5.3.

(2) Note that, by Proposition 5.4,

$$\mathbf{d} \in D_m^{\mathscr{D}} \quad \text{iff} \quad \text{either } \mathbf{d} = \mathbf{n} \text{ or } \mathbf{d} = \mathbf{o} \text{ or } \mathbf{d} = \mathbf{0}_m.$$

Assume that $\mathbf{c}, \mathbf{d} \in D_m^{\mathscr{D}}$. There are the following cases to consider:

(a) $\mathbf{c} = \mathbf{d}$. Observe that $\mathbf{c} \sqcup_m \mathbf{d} = \mathbf{c}$. Thus, $\mathbf{c} \sqcup_m \mathbf{d} \in D_m^{\mathscr{D}}$.

(b) Either $\mathbf{c} = \mathbf{0}_m$ or $\mathbf{d} = \mathbf{0}_m$. Then, by Exercise 5.10, $\mathbf{c} \sqcup_m \mathbf{d} = \mathbf{0}_m$. Thus, $\mathbf{c} \sqcup_m \mathbf{d} \in D_m^{\mathscr{D}}$.

(c) Either $\mathbf{c}$ is $\mathbf{o}$ and $\mathbf{d}$ is $\mathbf{n}$ or vice-versa. Observe that $\mathbf{o} \sqcup_m \mathbf{n} = \mathbf{0}_m$ by Exercise 5.10. Thus, $\mathbf{c} \sqcup_m \mathbf{d} \in D_m^{\mathscr{D}}$.

(3) Assume that $\mathbf{c} \leq_m \mathbf{d}$. Moreover, $\mathbf{d} \leq_m \mathbf{d}$ and so $\mathbf{d}$ is an upper bound of $\{\mathbf{c}, \mathbf{d}\}$. Since $\mathbf{c} \sqcup_m \mathbf{d}$ is a least upper bound of $\{\mathbf{c}, \mathbf{d}\}$, then $\mathbf{c} \sqcup_m \mathbf{d} \leq_m \mathbf{d}$. On the other hand, $\mathbf{d} \leq_m \mathbf{c} \sqcup_m \mathbf{d}$. Hence, by antisymmetry of $\leq_m$ in D_m, the result follows. $\square$

5.3 Many-to-One Completeness

Starting with the important notion of many to one completeness, we now take the opportunity to carry out a deeper study of the listable sets in general and of the listable non-decidable sets in particular.

Definition 5.6. Given $D \in \mathscr{S}_1$, we say that D is *many-to-one complete* or m-*complete* if:

- $D \in \mathscr{L}_1$;

- $C \leq_m D$ for each $C \in \mathscr{L}_1$.

Using Proposition 5.1, we obtain the following properties of m-complete sets.

Proposition 5.9. Every m-complete set is listable but non-decidable.

Proof. Let D be m-complete and, so, by definition, listable and such that every listable unary set is m-reducible to it. Thus, in particular,

$$\mathbb{K} \leq_m D$$

since $\mathbb{K}$ is a unary listable set. Assume, by contradiction, that D is decidable. Hence, by Proposition 5.1, $\mathbb{K}$ is also decidable contradicting Proposition 3.6. $\qquad \square$

Proposition 5.10 (Many-to-One Completeness Criteria). The following assertions provide necessary and sufficient conditions for m-completeness:

1. For any proper universal function u:

    (a) D is m-complete if and only if $\mathbb{K}^u \equiv_m D$;

    (b) D is m-complete if and only if $D \in \mathscr{L}_1$ and $\mathbb{K}^u \leq_m D$.

2. D is m-complete if and only if $D \in \mathbf{0}'_m$.

Proof. (1a) For the left to right implication, let D be m-complete. Then, $\mathbb{K}^u \leq_m D$ since $\mathbb{K}^u$ is a listable set. Moreover, $D \leq_m \mathbb{K}^u$, thanks to Proposition 5.3 and the listability of D.

For the right to the left implication observe that if $\mathbb{K}^u \equiv_m D$ then $D \leq_m \mathbb{K}^u$. Thus, by Proposition 5.3, D is a listable set. Moreover, if $\mathbb{K}^u \equiv_m D$ then $\mathbb{K}^u \leq_m D$ and, so, given an arbitrary listable set C, since, again by Proposition 5.3, $C \leq_m \mathbb{K}^u$, we obtain $C \leq_m D$ by transitivity (Exercise 5.1).

(1b) Note that, for any proper universal function u, thanks again to Proposition 5.3, $D \in \mathscr{L}_1$ if and only if $D \leq_m \mathbb{K}^u$. Thus, 1(b) is an immediate corollary of (1a).

(2) Observe that $D \in \mathbf{0}'_m$ if and only if $\mathbb{K} \equiv_m D$. Hence, since univ is a proper universal function, the result follows as an instance of (1a). $\qquad \square$

Exercise 5.11. Show that $\mathbb{K}$ is m-complete.

Example 5.1. We prove that $D = \{p \in \mathbb{N} : c \in \text{dom } \phi_p\}$ is an m-complete set for each $c \in \mathbb{N}$. By Proposition 5.10, for showing that D is m-complete it is enough to verify that D is listable and $\mathbb{K} \leq_m D$.

We start by showing that D is listable. Let

$$A = \{f \in \mathscr{C}_1 : c \in \text{dom } f\} \text{ and } P = \{p \in \mathbb{N} : c \in \text{dom } \theta_p\}.$$

We leave to the interested reader the proof that P is listable (in fact, P is decidable). We now prove that the condition of the Rice-Shapiro-McNaughton-Myhill Theorem (Proposition 4.8) holds:

($\subseteq$) Assume that $f \in A$. Then $c \in \text{dom } f$. Let ϑ be the finite function such that $\text{dom } \vartheta = \{c\}$ and $\vartheta(c) = f(c)$. Then, $\vartheta \subseteq f$. Furthermore there is $p \in \mathbb{N}$ such that $\vartheta = \theta_p$. By definition of P, we have $p \in P$ and, so, $f \in \bigcup_{p \in P}\{f : \theta_p \subseteq f\}$.

($\supseteq$) Conversely, assume that $f \in \bigcup_{p \in P}\{f : \theta_p \subseteq f\}$. Then, there is $p \in P$ such that $\theta_p \subseteq f$. By definition of P, $c \in \text{dom } \theta_p \subseteq \text{dom } f$. Hence, $f \in A$. So D is listable.

We now prove that $\mathbb{K} \leq_m D$. To this end, we consider the following function

$$v = \lambda\, p, x. \begin{cases} x & \text{if } p \in \mathbb{K} \\ \text{undefined} & \text{otherwise} \end{cases}.$$

Function v is clearly computable. Hence, by the *s-m-n* property there is $s \in \text{m}\mathscr{C}_1$ such that

$$v_p = \phi_{s(p)}$$

for every $p \in \mathbb{N}$. Then,

- if $p \in \mathbb{K}$ then $\phi_{s(p)} = v_p = \lambda x.x$ and, so, $\text{dom } \phi_{s(p)} = \mathbb{N}$ which implies that $c \in \text{dom } \phi_{s(p)}$ and, consequently, $s(p) \in D$;

- if $p \notin \mathbb{K}$ then $\phi_{s(p)} = v_p = \lambda x.\,\text{undefined}$ and, so, $\text{dom } \phi_{s(p)} = \emptyset$ which implies that $c \notin \text{dom } \phi_{s(p)}$ and, consequently, $s(p) \notin D$;

for every $p \in \mathbb{N}$. Hence,

$$p \in \mathbb{K} \quad \text{iff} \quad s(p) \in D$$

for every $p \in \mathbb{N}$. So,

$$\mathbb{K} \leq_m D.$$

Thus, by Propostion 5.10, we conclude that D is m-complete.

The reader may wonder if the class of listable non-decidable unary sets coincides with the class of m-complete sets. Reaching the negative answer to this question, following an original idea by Emile Post, is one of the main objectives of this chapter.

Observe that, so far, we only proved that $D_m^{\mathscr{L}} \setminus D_m^{\mathscr{D}}$ is countable (see Exercise 5.9, Proposition 5.4 and Proposition 5.6) and contains at least one element (namely, $0'_m$, see Proposition 5.5). The fact that $D_m^{\mathscr{L}} \setminus D_m^{\mathscr{D}}$ is not a singleton follows from the Many-to-One Completeness Criteria (Proposition 5.10) and from the existence (see Proposition 5.21) of listable non-decidable unary sets that are not m-complete. In fact, this set of degrees is denumerable but the proof of this result is beyond the scope of an introductory textbook.

5.4 Effective Non-Listability

Always within $\mathscr{S}_1$, we continue the study of the listable non-decidable sets with a closer look at their non-listable complements. From now on, we use the indexing of the listable unary sets provided by $\mathrm{Univ} = \mathrm{dom\,univ}$ in Chapter 3. Clearly, thanks to the fact that proper universal sets are isomorphic (see Exercise 4.18), the following concepts and results are robust in the sense that they do not depend on this choice.

Remark 5.6. Given $C, D \in \mathscr{S}_1$, we denote by

$$C \triangle D$$

the symmetric difference

$$(C \cap D^c) \cup (C^c \cap D) = (C \cup D) \setminus (C \cap D)$$

of sets C and D.

Clearly, if $C \subseteq D$ then $C \triangle D = D \setminus C$.

Definition 5.7. A unary set C is said to be *effectively non-listable* if there exists $s \in m\mathscr{C}_1$ such that, for every $p \in \mathbb{N}$,

$$s(p) \in (C \triangle \mathrm{Univ}_p).$$

In this case, the map s is said to be a *witness map* for the effective non-listability of C.

Intuitively, C is effectively non-listable if there is an algorithm s providing for each listable set Univ_p a point $s(p)$ where it differs from C.

Exercise 5.12. Show that if a set is effectively non-listable then it is non-listable.

The converse of the statement in Exercise 5.12 is false, as we shall see later on.

Proposition 5.11. The set $\mathbb{K}^c$ is effectively non-listable.

Proof. Take s as $id_\mathbb{N}$. We must prove that

$$s(p) = p \in \mathbb{K}^c \triangle \text{Univ}_p$$

for every $p \in \mathbb{N}$. That is, for every $p \in \mathbb{N}$,

$$p \in (\mathbb{K}^c \cap (\text{Univ}_p)^c) \cup (\mathbb{K} \cap \text{Univ}_p)$$

Let $p \in \mathbb{N}$. There are two cases:

(1) $p \in \mathbb{K}$. Then $p \in \text{dom}\,\lambda\,p.\text{univ}(p,p)$. Thus, $(p,p) \in \text{dom}\,\text{univ}$ and so $(p,p) \in \text{Univ}$. Therefore, $p \in \text{Univ}_p$. Thus, $p \in \mathbb{K} \cap \text{Univ}_p$.

(2) $p \in \mathbb{K}^c$. Then, $p \notin \mathbb{K}$. Hence, $p \notin \text{dom}\,\lambda\,p.\text{univ}(p,p)$. Thus, $(p,p) \notin \text{dom}\,\text{univ}$ and so $(p,p) \notin \text{Univ}$. Therefore, $p \notin \text{Univ}_p$ and so $p \in (\text{Univ}_p)^c$. Thus, $p \in \mathbb{K}^c \cap (\text{Univ}_p)^c$. $\square$

We now show that $\leq_m$, besides preserving non-listability (proved in Proposition 5.1), preserves also effective non-listability.

Proposition 5.12. Assume that $C \leq_m D$. Then, the effective non-listability of C entails the effective non-listability of D.

Proof. Let s' be a witness map for the effective non-listability of C and $s'' : C \leq_m D$. That is,

$$(\dagger) \quad s'(p) \in (C \triangle \text{Univ}_p)$$

for every $p \in \mathbb{N}$ and

$$(\dagger\dagger) \quad x \in C \quad \text{iff} \quad s''(x) \in D$$

for every $x \in \mathbb{N}$. Using the fact that Univ is a proper universal set and by Exercise 3.14, there is a computable unary map r such that

$$(\dagger\dagger\dagger) \quad (s'')^{-1}(\text{Univ}_p) = \text{Univ}_{r(p)}$$

for every $p \in \mathbb{N}$. Take

$$s = s'' \circ s' \circ r.$$

Then, as shown below, we have:

$$s(p) \in (D \triangle \text{Univ}_p)$$

for every $p \in \mathbb{N}$ and, so, D is effectively non-listable. Indeed, for any $p \in \mathbb{N}$, by $(\dagger)$, we have:

$$s'(r(p)) \in (C \triangle \text{Univ}_{r(p)})$$

and, so, by $(\dagger\dagger\dagger)$,

$$s'(r(p)) \in \left(C \triangle (s'')^{-1}(\text{Univ}_p)\right).$$

Hence,

$$\text{either} \qquad (*) \qquad s'(r(p)) \in \left(C \cap ((s'')^{-1}(\text{Univ}_p))^{\text{c}} \right)$$

$$\text{or} \qquad (**) \qquad s'(r(p)) \in \left(C^{\text{c}} \cap (s'')^{-1}(\text{Univ}_p) \right).$$

Observe that if $(*)$ holds then:

$$\begin{cases} s'(r(p)) \in C \text{ and, so, by } (\dagger\dagger), \ s''(s'(r(p))) \in D; \\ s'(r(p)) \notin (s'')^{-1}(\text{Univ}_p) \text{ and, so, } s''(s'(r(p))) \notin \text{Univ}_p. \end{cases}$$

Therefore, in case $(*)$,

$$s''(s'(r(p))) \in \left(D \cap \text{Univ}_p^{\text{c}} \right).$$

Similarly, using again $(\dagger\dagger)$, if $(**)$ holds then

$$s''(s'(r(p))) \in \left(D^{\text{c}} \cap \text{Univ}_p \right).$$

In conclusion, from the disjunction of $(*)$ and $(**)$ we reach

$$s''(s'(r(p))) \in \left((D \cap \text{Univ}_p^{\text{c}}) \cup (D^{\text{c}} \cap \text{Univ}_p) \right) = (D \triangle \text{Univ}_p)$$

as envisaged. □

We now generalize Proposition 5.11 to any proper universal function besides univ.

Proposition 5.13. Let u be a proper universal function. Then, $(\mathbb{K}^u)^{\text{c}}$ is an effectively non-listable set.

Proof. Observe that, by Proposition 5.11, $\mathbb{K}^{\text{c}}$ is an effectively non-listable set. Since, by Exercise 5.4, $\mathbb{K}^{\text{c}} \equiv_{\mathsf{m}} (\mathbb{K}^u)^{\text{c}}$, then by Proposition 5.12, we can conclude that $(\mathbb{K}^u)^{\text{c}}$ is also an effectively non-listable set. □

The following result provides a sufficient condition for a set of indices to be effectively non-listable.

Proposition 5.14. Let $A \subseteq \mathscr{C}_1$. Then ind$(A)$ is effectively non-listable whenever $\lambda x.\text{undefined} \in A$ and $A \neq \mathscr{C}_1$.

Proof. Let $f \in \mathscr{C}_1$ and $A \subseteq \mathscr{C}_1$ be such that $\lambda x.\text{undefined} \in A$ and $f \notin A$. Consider

$$v = \lambda p,x. \begin{cases} f(x) & \text{if } p \in \mathbb{K} \\ \text{undefined} & \text{otherwise} \end{cases}.$$

Clearly, v is computable and so, by the *s-m-n* property of univ, there is $s \in \mathsf{m}\mathscr{C}_1$ such that

$$v_p = \phi_{s(p)}$$

for every $p \in \mathbb{N}$. We now show that

$$s : \mathbb{K}^c \leq_m \mathrm{ind}(A).$$

Indeed:

(1) Assume that $p \in \mathbb{K}^c$. Then $\phi_{s(p)} = v_p = \lambda x . \mathrm{undefined}$ and so $\phi_{s(p)} \in A$. Hence, $s(p) \in \mathrm{ind}(A)$.

(2) Assume that $p \notin \mathbb{K}^c$. Then $p \in \mathbb{K}$ and so $\phi_{s(p)} = v_p = f$. Hence, $\phi_{s(p)} \notin A$ and, consequently $s(p) \notin \mathrm{ind}(A)$.

Thus, by Proposition 5.12 and Proposition 5.11 , $\mathrm{ind}(A)$ is effectively non-listable.   □

Example 5.2. For example to conclude that the set of indices of

$$A = \{ f \in \mathscr{C}_1 : f \text{ is an injective function} \}$$

is effectively non-listable, it is enough to observe that $\lambda x . \mathrm{undefined} \in A$ (since it is vacuously injective) and $\mathscr{C}_1 \neq A$ (for instance $\lambda x . 0 \notin A$). The thesis follows by Proposition 5.14.

Since $\mathbb{K}$ is m-complete (Exercise 5.11) and $\mathbb{K}^c$ is effectively non-listable (Proposition 5.11), it is natural to ask whether a listable set is m-complete if and only if its complement is effectively non-listable. We proceed to establish the positive answer to this question.

Proposition 5.15. If a set is m-complete then its complement is effectively non-listable.

Proof. Assume that C is m-complete. Then $\mathbb{K} \leq_m C$ since $\mathbb{K}$ is listable. Thus, $\mathbb{K}^c \leq_m C^c$ (Proposition 5.1). Since $\mathbb{K}^c$ is effectively non-listable (Proposition 5.11) then, by Proposition 5.12, C^c is effectively non-listable.   □

Furthermore, the converse of this proposition also holds for listable sets, in consequence of the following lemma.

Proposition 5.16. Let $C \in \mathscr{L}_1$ and D be effectively non-listable. Then, $C^c \leq_m D$.

Proof. Consider the binary set
$$V = C \times \mathbb{N}.$$

Clearly, V is listable thanks to the listability of C. Moreover,

- $p \in C$ implies $V_p = \mathbb{N}$;

- $p \notin C$ implies $V_p = \emptyset$.

Since Univ is a proper universal set, there is $s \in \mathrm{m}\mathscr{C}_1$ such that

$$V_p = \mathrm{Univ}_{s(p)}$$

for every $p \in \mathbb{N}$ and, so,

- $p \in C$ implies $\mathrm{Univ}_{s(p)} = \mathbb{N}$;
- $p \notin C$ implies $\mathrm{Univ}_{s(p)} = \emptyset$.

Let r be a witness map for the effective non-listability of D. Observe that

$$r(s(p)) \in (D \triangle \mathrm{Univ}_{s(p)})$$

for each $p \in \mathbb{N}$. We now show that

$$r \circ s : C^c \leq_m D.$$

Indeed,

- $p \in C^c$ implies $p \notin C$ that implies $r(s(p)) \in (D \triangle \mathrm{Univ}_{s(p)}) = (D \triangle \emptyset) = D$;
- $p \notin C^c$ implies $p \in C$ that implies $r(s(p)) \in (D \triangle \mathrm{Univ}_{s(p)}) = (D \triangle \mathbb{N}) = D^c$.

Therefore, $C^c \leq_m D$. □

Proposition 5.17. A listable set is m-complete if and only if its complement is effectively non-listable.

Proof. Proposition 5.15 provides the left to right implication of the thesis. Conversely, let E be a listable set such that E^c is effectively non-listable. Then, since $\mathbb{K}$ is listable, by Proposition 5.16 we have

$$\mathbb{K}^c \leq_m E^c$$

and, so,

$$\mathbb{K} \leq_m E$$

by Proposition 5.1. Therefore E is m-complete by Proposition 5.10. □

Exercise 5.13. Show that:

1. A unary set is effectively non-listable if and only if the complement of some m-complete set is m-reducible to it.

2. A unary set is effectively non-listable if and only if the complement of each m-complete set is m-reducible to it.

5.5 Post's Construction

We now proceed to show that there is a listable non-decidable set which is not m-complete, or, equivalently, that there is a listable non-decidable set with complement which is not effectively non-listable (Proposition 5.17). So, in the language of m-degrees, we prove that

$$\{\mathbf{0}'_m\} \subsetneq (D_m^{\mathscr{L}} \setminus D_m^{\mathscr{D}}).$$

To this end, it is convenient to enrich further our terminology concerning the lista-bility of sets.

Definition 5.8. A set is said to be *immune* if it is infinite but does not contain any infinite listable subset.

Clearly, no immune set is listable.

Definition 5.9. A set is said to be *simple* if it is listable and its complement is immune.

Obviously, every simple set is non-decidable.

Proposition 5.18 (Post's Construction). There exists a simple set.

Proof. Post's idea was to build a set with at least an element (but not too many ele-ments) of every infinite listable set. This objective can be achieved by taking the range of a particular computable unary function. Let f be a unary function such that:

$$\left\{ \begin{array}{l} (\dagger) \quad \text{Univ}_p \text{ is infinite implies } \begin{cases} p \in \text{dom} f \\ f(p) \in \text{Univ}_p \end{cases} \\ (\ddagger) \quad p \in \text{dom} f \text{ implies } f(p) > 2p \end{array} \right\}$$

for every $p \in \mathbb{N}$. Recall that $\text{Univ}_p = (\text{dom univ})_p = \text{dom univ}_p$ for each $p \in \mathbb{N}$. Con-sider the following program P:

```
function (p) (
    b = true;
    t = 2p+1;
    while b do (
        x = 2p+1;
        while x ≤ t ∧ b do
            if stceval(P_univ, p, x, t, −1) ≠ −1 then
                b = false
            else
                x = x+1;
        t = t+1
    );
    return x
)
```

where P_{univ} is a program that computes univ. Observe that function f computed by P is such that $f(p)$ is the least $x \geq 2p+1$ for which the execution of P_{univ} on (p,x) takes at most t units of time such that $t \geq x$ and the execution of P_{univ} on (p,x') takes $t' \geq t$ units of time for every $x < x' \leq t$. We now show that f fulfills the envisaged properties:

(1) f satisfies (†). Assume that Univ_p is infinite. Then $\text{dom}\,\text{univ}_p$ is also infinite. Hence there is an infinite number of $x > 2p$ such that the execution of P_{univ} on (p,x) terminates in a finite number of steps and let $x' > 2p$ be the smallest such natural number and t' its execution time. Then, $f(p) = x'$ when the execution time of P_{univ} on (p,x'') is greater than or equal to t' for $x' < x'' \leq t'$. Otherwise, $f(p)$ is the least x'' with this property.

(2) f satisfies (‡) by construction.

It remains to show that range f is simple. Indeed:

(1) range f is listable. Immediate by the Listability Criteria (Proposition 2.11) since f is computable.

(2) $(\text{range } f)^{\complement}$ is infinite. For each $k \in \mathbb{N}$, the elements of range f contained in the set

$$\{y \in \mathbb{N} : y \leq 2k\},$$

that is, the elements of the set

$$\{f(p) : p \in \text{dom}\, f \text{ and } f(p) \leq 2k\},$$

are in the set

$$\{f(p) : p \in \text{dom}\, f \text{ and } p < k\}$$

because $f(p) > 2p$ for every $p \in \text{dom}\, f$, thanks to (‡). As a consequence, range f contains at most k elements in the set $\{y \in \mathbb{N} : y \leq 2k\}$. Therefore, for each $k \in \mathbb{N}$, $(\text{range } f)^{\complement}$ contains more than k elements in $\{y \in \mathbb{N} : y \leq 2k\}$. Hence, the number of elements of $(\text{range } f)^{\complement}$ is greater than k for each $k \in \mathbb{N}$. Thus, $(\text{range } f)^{\complement}$ is infinite.

(3) No infinite listable set is a subset of $(\text{range } f)^{\complement}$. Assume, by contradiction, that there is an infinite and listable set $C \subseteq (\text{range } f)^{\complement}$. Then, for some $p \in \mathbb{N}$, $C = \text{Univ}_p$. Thanks to (†), $f(p) \in C$. Thus, $C \not\subseteq (\text{range } f)^{\complement}$ contradicting the hypothesis. □

The following exercises are used in the proof of Proposition 5.19. The first provides a decidable set which is universal for the class of all finite subsets of $\mathbb{N}$.

Exercise 5.14. Define $\Theta \in \mathscr{D}_2$ such that:

- $\{\Theta_p : p \in \mathbb{N}\}$ is the class of all finite subsets of $\mathbb{N}$;

- $\Theta_{p_1} = \Theta_{p_2}$ implies $p_1 = p_2$.

Hint: Adapt the ideas of the proof of Proposition 4.7.

Exercise 5.15. Representing each finite unary set by the list of its elements in increasing order, present an algorithm in for computing the map

$$\Psi_\Theta$$

that assigns to each finite subset D of $\mathbb{N}$ the unique $p \in \mathbb{N}$ such that $\Theta_p = D$.

Proposition 5.19. If a set is effectively non-listable then it is not immune.

Proof. Let Θ be the set defined in Exercise 5.14. Since Univ is a proper universal set and $\Theta \in \mathscr{L}_2$ (Θ is decidable), there is $r \in m\mathscr{C}_1$ such that

$$\Theta_p = \text{Univ}_{r(p)}$$

for every $p \in \mathbb{N}$. Let C be an effectively non-listable set with witness map s. That is:

$$s(p) \in (C \triangle \text{Univ}_p)$$

for every $p \in \mathbb{N}$. Clearly, C is necessarily infinite. However, C is not immune because it contains an infinite listable subset. Indeed, consider $h \in m\mathscr{F}_1$ defined inductively as follows:

- $h(0) = s(r(\Psi_\Theta(\emptyset)))$;

- $h(k+1) = s(r(\Psi_\Theta(\{h(0), \dots, h(k)\})))$.

Observe that, by construction of h, we have (the details of the proof are left to the reader):

- $\text{Univ}_{r(\Psi_\Theta(\emptyset))} = \Theta_{\Psi_\Theta(\emptyset)} = \emptyset$ and, so, by definition of s,

$$h(0) = s(r(\Psi_\Theta(\emptyset))) \in (C \triangle \emptyset) = C;$$

- $\text{Univ}_{r(\Psi_\Theta(\{h(0),\dots,h(k)\}))} = \Theta_{\Psi_\Theta(\{h(0),\dots,h(k)\})} = \{h(0), \dots, h(k)\}$ and, so,

$$h(k+1) = s(r(\Psi_\Theta(\{h(0), \dots, h(k)\}))) \in (C \triangle \{h(0), \dots, h(k)\});$$

- $(C \triangle \{h(0), \dots, h(k)\}) = (C \setminus \{h(0), \dots, h(k)\})$;

- $\text{range}\, h \subseteq C$;

- h is an injective map and, so, $\text{range}\, h$ is infinite;

- h is computable;

- $\text{range}\, h$ is listable (Proposition 2.11).

Therefore, C is not immune because it contains range h which is infinite and listable. □

Proposition 5.20. If a set is simple then it is not m-complete.

Proof. Immediate corollary of Propositions 5.17 and 5.19. □

Proposition 5.21. The set $D_m^{\mathscr{L}} \setminus D_m^{\mathscr{D}}$ is not a singleton.

Proof. Besides $\mathbf{0}'_m$, the set $D_m^{\mathscr{L}} \setminus D_m^{\mathscr{D}}$ contains at least the listable non-decidable degree of the simple set constructed in Proposition 5.18 which, by Proposition 5.20, is not m-complete and, thus, is not in $\mathbf{0}'_m$ thanks to Proposition 5.10. □

5.6 Productive Sets

One of the main objectives of the section is to prove another characterization of m-complete sets given by Myhill's Theorem. To this end, we start by introducing the concept of productive set.

Definition 5.10. A unary set C is said to be *productive* if there exists $f \in \mathscr{C}_1$ such that:

$$\text{Univ}_p \subseteq C \quad \text{implies} \quad \begin{cases} p \in \text{dom} f \\ f(p) \in (C \setminus \text{Univ}_p) \end{cases}$$

for every $p \in \mathbb{N}$. In this case, we say that function f is a *productive function* for C.

Proposition 5.22. Let $C \in \mathscr{S}_1$. Then, C is non-listable whenever C is a productive set.

Proof. Assume, by contradiction, that C is productive and listable. Let f be a productive function for C. Since C is listable, there is $p \in \mathbb{N}$ such that $\text{Univ}_p = C$. Hence, $p \in \text{dom} f$ and $f(p) \in (C \setminus \text{Univ}_p) = \emptyset$ which is an absurd. □

Proposition 5.23. Every effectively non-listable set is productive.

Proof. Assume that C is effectively non-listable. Then, there is $s \in m\mathscr{C}_1$ such that

$$s(p) \in (C \triangle \text{Univ}_p)$$

for every $p \in \mathbb{N}$. We now show that s is a productive function for C. Let $p \in \mathbb{N}$ and assume that $\text{Univ}_p \subseteq C$. Since s is a total map then $p \in \text{dom} s$. On the other hand, because $s(p) \in (C \triangle \text{Univ}_p)$ then either $s(p) \in C \cap (\text{Univ}_p)^c$ or $s(p) \in C^c \cap \text{Univ}_p$. But, since $\text{Univ}_p \subseteq C$ then $C^c \cap \text{Univ}_p = \emptyset$. Thus, $s(p) \in C \cap (\text{Univ}_p)^c$, that is, $s(p) \in (C \setminus \text{Univ}_p)$. Hence, C is productive. □

We proceed to show that productivity entails effective non-listability for complements of listable sets. To this end, we need the counterpart of Proposition 5.16 for productive sets.

Proposition 5.24. Let $C \in \mathcal{L}_1$ and D be productive. Then, $C^c \leq_m D$.

Proof. Let $f \in \mathcal{C}_1$ be a productive function for D. That is:

$$\text{Univ}_p \subseteq D \quad \text{implies} \quad \begin{cases} p \in \text{dom} f \\ f(p) \in (D \setminus \text{Univ}_p) \end{cases}$$

for every $p \in \mathbb{N}$. Consider the ternary set

$$V = \{(p,q,y) : p \in C \text{ and } y = f(q)\}.$$

For any $p, q \in \mathbb{N}$, denoting by $V_{(p,q)}$ the section of V at (p,q), we have:

- if $p \in C$ and $q \in \text{dom} f$ then $V_{(p,q)} = \{f(q)\}$;

- if $p \notin C$ or $q \notin \text{dom} f$ then $V_{(p,q)} = \emptyset$.

Since C is listable and f is computable, V is listable and so is the binary set

$$V' = \{(z,y) : K(z) \in C \text{ and } y = f(L(z))\}.$$

Clearly,

$$V_{(p,q)} = V'_{J(p,q)}$$

for every $p \in \mathbb{N}$. Since Univ is a proper universal set, there is $r \in \text{m}\mathcal{C}_1$ such that

$$V'_z = \text{Univ}_{r(z)}$$

for every $z \in \mathbb{N}$. Thus, in particular, we have:

$$V_{(p,q)} = \text{Univ}_{r(J(p,q))}$$

for every $p, q \in \mathbb{N}$. Thus, for any $p, q \in \mathbb{N}$, we have:

- if $p \in C$ and $q \in \text{dom} f$ then $\text{Univ}_{r(J(p,q))} = \{f(q)\}$;

- if $p \notin C$ or $q \notin \text{dom} f$ then $\text{Univ}_{r(J(p,q))} = \emptyset$.

Since $r \circ J \in \text{m}\mathcal{C}_2$, by Proposition 4.23, there is $\varphi \in \text{m}\mathcal{C}_1$ such that

$$\text{Univ}_{r(J(p,\varphi(p)))} = \text{Univ}_{\varphi(p)}$$

for every $p \in \mathbb{N}$. Therefore, for each $p \in \mathbb{N}$, choosing $q = \varphi(p)$, we obtain:

- if $p \in C$ and $\varphi(p) \in \operatorname{dom} f$ then $\operatorname{Univ}_{\varphi(p)} = \{f(\varphi(p))\}$;

- if $p \notin C$ or $\varphi(p) \notin \operatorname{dom} f$ then $\operatorname{Univ}_{\varphi(p)} = \emptyset$.

Observe that

$$\varphi(p) \in \operatorname{dom} f$$

for every $p \in \mathbb{N}$ as we show now by contradiction. Assume that $\varphi(p) \notin \operatorname{dom} f$. Then $\operatorname{Univ}_{\varphi(p)}$ is the empty set. Thus, $\operatorname{Univ}_{\varphi(p)}$ is a subset of D and, so, by the productivity of f for D, $\varphi(p)$ is in $\operatorname{dom} f$, contradicting the initial assumption. In short, $f \circ \varphi \in \operatorname{m}\mathscr{C}_1$ and, for any $p \in \mathbb{N}$, the following assertions hold:

- if $p \in C$ then $\operatorname{Univ}_{\varphi(p)} = \{f(\varphi(p))\}$;

- if $p \notin C$ then $\operatorname{Univ}_{\varphi(p)} = \emptyset$.

Observe that, by the productivity of f for D, we know that

$$\operatorname{Univ}_{\varphi(p)} \subseteq D \quad \text{implies} \quad f(\varphi(p)) \in (D \setminus \operatorname{Univ}_{\varphi(p)})$$

for every $p \in \mathbb{N}$. Therefore, we conclude that

- if $\operatorname{Univ}_{\varphi(p)} = \{f(\varphi(p))\}$ then $\operatorname{Univ}_{\varphi(p)} = \{f(\varphi(p))\} \not\subseteq D$, since otherwise we have $f(\varphi(p)) \notin (D \setminus \{f(\varphi(p))\})$. So $f(\varphi(p)) \notin D$;

- if $\operatorname{Univ}_{\varphi(p)} = \emptyset$ then $f(\varphi(p)) \in D$.

Therefore,

- if $p \in C$ then $f(\varphi(p)) \notin D$;

- if $p \notin C$ then $f(\varphi(p)) \in D$.

That is,

- if $p \notin C^{\text{c}}$ then $f(\varphi(p)) \notin D$;

- if $p \in C^{\text{c}}$ then $f(\varphi(p)) \in D$;

and, so, $f \circ \varphi : C^{\text{c}} \leq_{\text{m}} D$. $\qquad\square$

The next result shows that there is a close relationship between productivity and effective non-listability.

Proposition 5.25. The complement of a listable set is productive if and only if it is effectively non-listable.

Proof. In Proposition 5.23, we already proved that effective non-listability entails productivity. Conversely, let E be a listable set and assume that E^c is productive. Then, since $\mathbb{K}$ is listable, using Proposition 5.24, we obtain

$$\mathbb{K}^c \leq_m E^c$$

and, so,

$$\mathbb{K} \leq_m E$$

by Proposition 5.1. Thus, E is m-complete by Proposition 5.10, and, in consequence, by Proposition 5.17, E^c is effectively non-listable. $\qquad\square$

We are ready to prove another characterization of m-complete sets.

Proposition 5.26. [Myhill's Theorem] A unary set is m-complete if and only if it is listable and its complement is productive.

Proof. Immediate corollary of Propositions 5.25 and 5.17. $\qquad\square$

5.7   Solved Problems and Exercises

Problem 5.1. Assume that $s : C_1 \leq_m C_2$. Verify that:

1. $s(C_1) \subseteq C_2$;

2. $\chi^p_{C_1} = \chi^p_{C_2} \circ s$;

3. $\chi_{C_1} = \chi_{C_2} \circ s$.

Solution (1) By hypothesis, for every $x \in \mathbb{N}$,

$$x \in C_1 \quad \text{iff} \quad s(x) \in C_2$$

for all $x \in \mathbb{N}$, and so $s(C_1) \subseteq C_2$.

(2) Observe that, for every $x \in \mathbb{N}$,

$$\chi^p_{C_1}(x) = 1 \text{ iff } x \in C_1 \quad \text{iff} \quad s(x) \in C_2 \text{ iff } \chi^p_{C_2}(s(x)) = 1 \text{ iff } (\chi^p_{C_2} \circ s)(x) = 1.$$

Hence, $\chi^p_{C_1} = \chi^p_{C_2} \circ s$.

(3) The proof is similar to (2). $\qquad\triangledown$

Problem 5.2. Let $A \in \mathcal{S}_1$. Assume that $s : C \leq_m D$. Show that

$$x \in C \setminus s^{-1}(A) \quad \text{implies} \quad s(x) \in D \setminus A.$$

Solution Assume that $x \in C \setminus s^{-1}(A)$. Then $s(x) \in D$ since $s : C \leq_m D$. Suppose, by contradiction, that $s(x) \in A$. Thus,

$$x = s^{-1}(s(x)) \in s^{-1}(A)$$

which contradicts the fact $x \notin s^{-1}(A)$. ∇

Problem 5.3. Show that

$$\{p \in \mathbb{N} : \phi_p = \lambda x.0\} \equiv_m \{p \in \mathbb{N} : \phi_p \text{ is total}\}.$$

Solution Let

$$D_1 = \{p \in \mathbb{N} : \phi_p = \lambda x.0\} \quad \text{and} \quad D_2 = \{p \in \mathbb{N} : \phi_p \text{ is total}\}.$$

$(D_1 \leq_m D_2)$ We must prove that there is a map $s \in m\mathscr{C}_1$ such that, for every $p \in \mathbb{N}$,

$$\phi_p = \lambda x.0 \quad \text{iff} \quad \phi_{s(p)} \text{ is total.} \qquad (\ddagger)$$

Thus, if

$$\phi_{s(p)} = \begin{cases} \lambda x.0 & \text{if } \phi_p = \lambda x.0 \\ \text{a non-total function} & \text{otherwise} \end{cases}$$

then we can conclude $(\ddagger)$. For instance, if

$$\phi_{s(p)}(x) = \begin{cases} 0 & \text{if } \phi_p(x) = 0 \\ \text{undefined} & \text{otherwise} \end{cases}$$

for every $x \in \mathbb{N}$, then we have $(\ddagger)$. Consider the following computable binary function v:

$$v(p,x) = \begin{cases} 0 & \text{if } \phi_p(x) = 0 \\ \text{undefined} & \text{otherwise} \end{cases}$$

for every $p,x \in \mathbb{N}$. Then, by the *s-m-n* property there is $s \in m\mathscr{C}_1$ satisfying $(\ddagger)$.

$(D_2 \leq_m D_1)$ We must prove that there is $s \in m\mathscr{C}_1$ such that, for every $p \in \mathbb{N}$,

$$\phi_p \text{ is total} \quad \text{iff} \quad \phi_{s(p)} = \lambda x.0. \qquad (\ddagger)$$

Observe that,

$$\phi_{s(p)}(x) = \begin{cases} 0 & \text{if } \phi_p(x)\downarrow \\ \text{undefined} & \text{otherwise} \end{cases}$$

for every $p,x \in \mathbb{N}$, is enough to guarantee that $(\ddagger)$ holds. Consider the computable binary function v:

$$v(p,x) = \begin{cases} 0 & \text{if } \phi_p(x)\downarrow \\ \text{undefined} & \text{otherwise} \end{cases}$$

for every $p,x \in \mathbb{N}$. Hence, by the *s-m-n* property there is $s \in m\mathscr{C}_1$ satisfying $(\ddagger)$. ∇

Problem 5.4. Show that the relation between m-degrees:

$$d_m(C) \leq_m d_m(D) \quad \text{if} \quad C \leq_m D$$

for $C, D \in \mathscr{S}_1$, is well defined.

Solution Let $C, C', D, D' \in \mathscr{S}_1$ be such that $d_m(C') = d_m(C)$ and $d_m(D') = d_m(D)$. We must prove that

$$\text{if } d_m(C) \leq_m d_m(D) \text{ then } d_m(C') \leq_m d_m(D').$$

First, observe that

$$[C]_m = d_m(C) = d_m(C') = [C']_m$$

and so $C \equiv_m C'$. By a similar argument, we may also conclude that $D \equiv_m D'$. Hence, $C' \leq_m C$ and $D \leq_m D'$. Thus, if $d_m(C) \leq_m d_m(D)$ then, by definition, $C \leq_m D$ and by transitivity of $\leq_m$, we also have $C' \leq_m D'$, which implies, that $d_m(C') \leq_m d_m(D')$. ∇

Problem 5.5. Let $c \in \mathbb{N}$. Show that the set $\{p \in \mathbb{N} : \phi_p(p) = c\}$ is m-complete.

Solution By Proposition 5.10, for showing that a set D is m-complete it is enough to verify that D is listable and $\mathbb{K} \leq_m D$.

Let $D = \{p \in \mathbb{N} : \phi_p(p) = c\}$. We leave for the interested reader the prove that D is listable. To prove that $\mathbb{K} \leq_m D$, consider the function

$$v = \lambda \, p, x . \begin{cases} c & \text{if } p \in \mathbb{K} \\ \text{undefined} & \text{otherwise} \end{cases}.$$

This function is clearly computable. By the *s-m-n* property, there is $s \in m\mathscr{C}_1$ such that

$$v_p = \phi_{s(p)}$$

for every $p \in \mathbb{N}$. Then,

- if $p \in \mathbb{K}$ then $\phi_{s(p)} = v_p = \lambda x . c$ and, so, $\phi_{s(p)}(s(p)) = c$ which implies that $s(p) \in D$;

- if $p \notin \mathbb{K}$ then $\phi_{s(p)} = v_p = \lambda x . \text{undefined}$ and, so, $\phi_{s(p)}(s(p)) \neq c$ which implies that $s(p) \notin D$;

for every $p \in \mathbb{N}$. Hence,

$$p \in \mathbb{K} \quad \text{iff} \quad s(p) \in D$$

for every $p \in \mathbb{N}$, and, so,

$$\mathbb{K} \leq_m D.$$

Thus, we conclude that D is m-complete. ∇

Problem 5.6. Let $u \in \mathscr{F}_2$. Characterize the domain of

$$\lambda x. \begin{cases} u(x,x) & x \neq 0 \\ \text{undefined} & \text{otherwise} \end{cases} : \mathbb{N} \rightharpoonup \mathbb{N}$$

and its complement, assuming that u is:

1. computable;

2. universal computable;

3. proper universal.

Solution Let

$$D = \operatorname{dom} \lambda x. \begin{cases} u(x,x) & x \neq 0 \\ \text{undefined} & \text{otherwise} \end{cases}.$$

(1) Assume that u is a computable function. Then the program

```
function (x) (
    if x ≠ 0 then
        return Pu(x,x)
    else
        while true do null
)
```

where P_u is a program that computes u, computes the given function. Hence, D is a listable set because it is the domain of a computable function (Proposition 2.11).

(2) If u is a universal computable function then D is a listable but not a decidable set. Indeed,

$$D = \mathbb{K}^u \setminus \{0\}$$

and, so,

$$D^{\complement} = (\mathbb{K}^u)^{\complement} \cup \{0\}.$$

We also know that $\mathbb{K}^u = \operatorname{dom}\operatorname{diag}^u$ is a listable but not a decidable set (Proposition 3.6). Hence, by Post's Theorem (Proposition 2.6), $(\mathbb{K}^u)^{\complement}$ is not a listable set. We have two cases to consider:

(a) If $0 \notin \mathbb{K}^u$ then $D = \mathbb{K}^u$ and, so, D is a listable but not a decidable set and $D^{\complement} = (\mathbb{K}^u)^{\complement}$ is not a listable set.

(b) Otherwise, $\mathbb{K}^u = D \cup \{0\}$ and, so, if D was a decidable set so was $\mathbb{K}^u$ (since the union of two decidable sets is still decidable by Exercise 2.2). Hence, D is not a decidable set. Moreover, by (1), D is listable. Therefore, $D^{\complement}$ is not a listable set by Post's Theorem (Proposition 2.6).

(3) If u is a proper universal function then by (2) D is non-decidable and listable. Moreover, D is m-complete. Indeed, D is m-complete since it is listable, and $\mathbb{K}^u \leq_m D$ as we shall see now. If $0 \notin \mathbb{K}^u$ then $D = \mathbb{K}^u$ and, so, the identity map reduces $\mathbb{K}^u$ to D. Otherwise, $\mathbb{K}^u = D \cup \{0\}$ and, choosing $d \in D$ (since $D \neq \emptyset$ because D is not a decidable set), the computable map

$$\lambda x. \begin{cases} x & \text{if } x \neq 0 \\ d & \text{otherwise} \end{cases}$$

reduces $\mathbb{K}^u$ to D. Therefore, D^c is effectively non-listable (Proposition 5.17) and productive (Proposition 5.23). Furthermore, D^c is not immune (Proposition 5.19). ∇

Problem 5.7. Verify that $\mathbb{K}^c$ is productive, with productive function $\lambda\, p.p$.

Solution Let $f = \lambda\, p.p$ and $q \in \mathbb{N}$. Assume that $\Gamma_q \subseteq \mathbb{K}^c$. Clearly, $q \in \operatorname{dom} f$. Hence, we only have to check that $f(q) = q \in (\mathbb{K}^c \setminus \Gamma_q)$. If $q \notin \mathbb{K}^c$ then $\operatorname{univ}(q,q)$ is defined and so $q \in \Gamma_q = \operatorname{dom\,univ}_q$, that is clearly impossible because $\Gamma_q \subseteq \mathbb{K}^c$. Hence, $q \in \mathbb{K}^c$. In this case, $\operatorname{univ}(q,q)$ is not defined and, so, $q \notin \Gamma_q$. From this, we conclude that $q \in (\mathbb{K}^c \setminus \Gamma_q)$. Thus, $\mathbb{K}^c$ is productive. ∇

Problem 5.8. Let u be a proper universal function. Classify the set $\{p \in \mathbb{N} : \phi_p^u \neq \lambda x.x + 1\}$.

Solution Observe that $(\mathbb{K}^u)^c$ is an effectively non-listable set, by Proposition 5.13. Thus if we show that

$$(\mathbb{K}^u)^c \leq_m \{p \in \mathbb{N} : \phi_p^u \neq \lambda x.x + 1\} \qquad (\ddagger)$$

then we can use Proposition 5.12 to conclude that $\{p \in \mathbb{N} : \phi_p^u \neq \lambda x.x + 1\}$ is an effectively non-listable set. Hence, we must find a map $s \in \mathrm{m}\mathscr{C}_1$ such that for every $p \in \mathbb{N}$,

$$p \in (\mathbb{K}^u)^c \quad \text{iff} \quad s(p) \in \{p \in \mathbb{N} : \phi_p^u \neq \lambda x.x + 1\},$$

that is

$$p \in (\mathbb{K}^u)^c \quad \text{iff} \quad \phi_{s(p)}^u \neq \lambda x.x + 1$$

which is equivalent to

$$p \in \mathbb{K}^u \quad \text{iff} \quad \phi_{s(p)}^u = \lambda x.x + 1.$$

Therefore, if

$$\phi_{s(p)} = \begin{cases} \lambda x.x + 1 & \text{if } p \in \mathbb{K}^u \\ \lambda x.\text{undefined} & \text{otherwise} \end{cases}$$

we can conclude ($\ddagger$). Consider the following binary function v:

$$v_p = \begin{cases} \lambda x . x + 1 & \text{if } p \in \mathbb{K}^u \\ \lambda x . \text{undefined} & \text{otherwise} \end{cases},$$

that is,

$$v(p,x) = \begin{cases} x + 1 & \text{if } p \in \mathbb{K}^u \\ \text{undefined} & \text{otherwise} \end{cases}.$$

Since

$$v = (\lambda u, w . u \times (w+1)) \circ \langle \chi_{\mathbb{K}^u}^{p} \circ \text{proj}_{[1]}^{\mathbb{N}^2}, \text{proj}_{[2]}^{\mathbb{N}^2} \rangle$$

we can conclude that v is a computable function. Thus, by the *s-m-n* property there is $s \in \mathrm{m}\mathscr{C}_1$ fulfilling ($\ddagger$). Therefore,

$$\{p \in \mathbb{N} : \phi_p^u \neq \lambda x . x + 1\}$$

is an effectively non-listable set, so it is not a listable set and à fortiori is not a decidable set. On the other hand, by Proposition 5.23, it is a productive set. Moreover, by Proposition 5.19, it is not a immune set.

Furthermore, its complement $\{p \in \mathbb{N} : \phi_p^u = \lambda x . x + 1\}$, is not a listable set, by Exercise 4.26 and so it is not m-complete by definition. $\qquad \nabla$

Supplementary Exercises

Exercise 5.16. Define degree complementation as follows

$$\mathbf{d}^c = [D^c]_\mathrm{m} \text{ where } D \in \mathbf{d}.$$

Is it well defined? Investigate its properties.

Exercise 5.17. Show that the following sets are effectively non-listable:

- $\{p \in \mathbb{N} : \phi_p \neq \lambda x . 0\}$;
- $\{p \in \mathbb{N} : c \notin \text{dom } \phi_p\}$.

Exercise 5.18. Show that the following sets are m-complete.

- $\{p \in \mathbb{N} : p \in \text{range } \phi_p\}$;
- $\{p \in \mathbb{N} : \phi_p \text{ is not injective}\}$.

Exercise 5.19. Show that if B is a decidable set and $A \cap B$ is a productive set then A is also a productive set.

Exercise 5.20. Given a proper universal set U and $C \in \mathscr{S}_1$, we say that C is U-*effectively non-listable* if there exists $s \in m\mathscr{C}_1$ such that, for every $p \in \mathbb{N}$, $s(p) \in (C \triangle \Gamma_p^U)$. Let U and U' be proper universal sets. Show that C is U-effectively non-listable if and only if C is U'-effectively non-listable.

Exercise 5.21. Set up and study the basic properties of the category mS_1 of unary sets and m-reductions.

Exercise 5.22. Let mL_1 be the full subcategory of mS_1 composed of the listable unary sets and their m-reductions. Characterize the m-complete sets in mL_1.

Chapter 6

Operators

In Chapter 3 we discussed how some operations on computable functions (like composition) could be lifted to computable maps on their indices. Furthermore, we also investigated how to lift to indices the notion of computability of sequences of computable functions and of sequences of listable sets. This issue is the main goal of this chapter. We start by defining computability with oracles.

6.1 Computing with Oracles

By an *(evaluation) oracle* for $f \in \mathscr{F}_1$ we understand an entity that on receiving x provides the value $f(x)$ in a finite amount of time whenever $x \in \text{dom} f$, and is undefined otherwise. Observe that if f is computable then there is a program P_f that computes f and behaves exactly in the same way as an oracle for f. In this case, the references in a program P to an oracle for f can be replaced by references to P_f, without changing the behaviour of P.

As it might be expected, it is necessary to enrich the computational model with oracles. We adapt the approach described in [52] and [17] for Turing machines and register machines, respectively. To this end, for each $f \in \mathscr{F}_1$, we use the expression $\text{Oracle}(f)$ for representing an invocation to an oracle for f. The objective is to define the notion of computable function relatively to a finite set of functions (oracles).

Definition 6.1. Given $f \in \mathscr{F}_1$, the evaluation of

$$\text{Oracle}(f)(x)$$

in the adopted computational model enriched with an oracle for f will return $f(x)$ when $x \in \text{dom} f$, otherwise, the evaluation does not provide an answer in a finite amount of time.

Definition 6.2. Let $F \subseteq \mathscr{F}_1$ be a finite set. We say that $g \in \mathscr{F}_1$ is *computable relatively to F*, in short,

$$computable/F$$

if there is a program that computes g in the adopted computational model enriched with an oracle for each $f \in F$.

Exercise 6.1. Show that $\chi_\mathbb{K}$ is computable relatively to $\{\chi_{\mathbb{K}^c}^p\}$.

Exercise 6.2. Let $F \subseteq \mathscr{F}_1$ be a finite set, $f, g \in \mathscr{F}_1$ and $f \notin F$. Let P_g be a program that computes g relatively to $F \cup \{f\}$. Assume that $\mathsf{Oracle}(f)$ does not occur in P_g. Show that P_g is a program that computes g relatively to F.

Remark 6.1. Given a finite set $F \subseteq \mathscr{F}_1$, we denote by

$$\mathscr{C}_1/F$$

the set $\{g \in \mathscr{F}_1 : g \text{ is computable}/F\}$ and by

$$\mathsf{m}\mathscr{C}_1/F$$

the set $\{g \in \mathscr{C}_1/F : g \text{ is total}\}$. Moreover, when $F = \{f_1, \dots, f_k\}$ we may use $f_1, \dots, f_k$ instead of F.

Exercise 6.3. Given finite sets $F, G \subseteq \mathscr{F}_1$, show that:

1. $F \subseteq \mathscr{C}_1/F$;

2. $\mathscr{C}_1 \subseteq \mathscr{C}_1/F$;

3. If $F \subseteq \mathscr{C}_1$ then $\mathscr{C}_1 = \mathscr{C}_1/F$;

4. if $F \subseteq G$ then $\mathscr{C}_1/F \subseteq \mathscr{C}_1/G$;

5. if $G \subseteq \mathscr{C}_1/F$ then $\mathscr{C}_1/G \subseteq \mathscr{C}_1/F$.

6.2   Properties of Operators

Herein we concentrate on maps between unary functions over the set of natural numbers. We analyze different properties of operators as well as their relationship. Furthermore, we provide a characterization of computable operators, including their counterparts at the level of the indices.

Definition 6.3. An *operator* is a map $\Psi : \mathscr{F}_1^n \to \mathscr{F}_1$ for some $n \in \mathbb{N}^+$.

Example 6.1. The composition of unary functions can be seen as a binary operator

$$\circ : \mathscr{F}_1 \times \mathscr{F}_1 \to \mathscr{F}_1$$

such that $\circ(f,g) = g \circ f$ for every $f, g \in \mathscr{F}_1$.

In the following we concentrate on operators from $\mathscr{F}_1$ to $\mathscr{F}_1$ and leave as an exercise the study of operators with other sources. Accordingly, from now on, unless otherwise explicitly stated, by an *operator* we mean a map from $\mathscr{F}_1$ to $\mathscr{F}_1$.

Remark 6.2. Given an operator Ψ, a unary function f and $x \in \mathbb{N}$, it is usual to write

$$\Psi(f)(x)$$

for $(\Psi(f))(x)$.

Definition 6.4. An operator $\Psi : \mathscr{F}_1 \to \mathscr{F}_1$ is said to be *computable* if there is a program P_Ψ such that, for every $f : \mathbb{N} \rightharpoonup \mathbb{N}$ the program

$$\mathsf{P}_\Psi(\mathsf{Oracle}(f))$$

computes $\Psi(f)$ relatively to f. In this situation, we say that program P_Ψ *computes* operator Ψ.

Example 6.2. Consider the operator

$$\Psi = \lambda f . (\lambda x . f(x+1)) : \mathscr{F}_1 \to \mathscr{F}_1.$$

The following program P_Ψ

```
function (o) (
    return function (x) (return o(x + 1))
)
```

computes Ψ as we now show. Let $f \in \mathscr{F}_1$. We must prove that $\mathsf{P}_\Psi(\mathsf{Oracle}(f))$ computes $\Psi(f)$ relatively to f.

(1) Assume that $x \in \operatorname{dom}\Psi(f)$. Then $x+1 \in \operatorname{dom} f$. So the execution, in the adopted computational model enriched with an oracle for f, of $\mathsf{Oracle}(f)(x+1)$ returns $f(x+1)$ in a finite amount of time. Therefore, the execution, in the adopted computational model enriched with an oracle for f, of $\mathsf{P}_\Psi(\mathsf{Oracle}(f))(x)$ returns $f(x+1)$ in a finite amount of time.

(2) Assume that $x \notin \operatorname{dom}\Psi(f)$. Then $x+1 \notin \operatorname{dom} f$. So the execution of $\mathsf{Oracle}(f)(x+1)$, in the adopted computational model enriched with an oracle for f, does not terminate. Therefore, the execution, in the adopted computational model enriched with an oracle for f, of $\mathsf{P}_\Psi(\mathsf{Oracle}(f))(x)$ does not terminate.

Proposition 6.1. Every computable operator maps computable functions to computable functions.

Proof. Let Ψ be computed by program P_Ψ and $f \in \mathscr{C}_1$. Then, $P_\Psi(\text{Oracle}(f))$ computes $\Psi(f)$ relatively to f. Let P_f be a program such that $\text{Oracle}(f)$ and P_f return the same value on input x when the execution of $\text{Oracle}(f)$ on x terminates. Let $\tilde{P}_\Psi$ be the program obtained from P_Ψ by replacing the expression $\text{Oracle}(f)$ by P_f. Then $\tilde{P}_\Psi(P_f)$ and $P_\Psi(\text{Oracle}(f))$ return the same value on $x \in \mathbb{N}$. Observe that $\text{Oracle}(f)$ does not occur in $\tilde{P}_\Psi(P_f)$. Hence, by Exercise 6.2, $\tilde{P}_\Psi(P_f)$ computes $\Psi(f)$.   □

Our objective at this stage is to characterize other interesting subclasses of the class of operators.

Definition 6.5. An operator Ψ is said to be *monotonic* if, for every $f, g \in \mathscr{F}_1$,

$$\Psi(f) \subseteq \Psi(g)$$

whenever $f \subseteq g$.

Example 6.3. Consider the operator:

$$\Psi = \lambda f. \begin{cases} \lambda x.1 & \text{if } \text{dom} f = \mathbb{N} \\ \lambda x.\text{undefined} & \text{otherwise} \end{cases}.$$

We show that Ψ is monotonic. Let $f \subseteq g$. We have to show that $\Psi(f) \subseteq \Psi(g)$. We need to consider three cases:

(1) $\text{dom} f \subseteq \text{dom} g \subsetneq \mathbb{N}$:

$$\Psi(f) = \lambda x.\text{undefined} = \Psi(g).$$

(2) $\text{dom} f \subsetneq \text{dom} g = \mathbb{N}$:

$$\Psi(f) = \lambda x.\text{undefined} \subseteq \lambda x.1 = \Psi(g).$$

(3) $\text{dom} f = \text{dom} g = \mathbb{N}$:

$$\Psi(f) = \lambda x.1 = \Psi(g).$$

Hence, Ψ is an example of a monotonic operator.

Exercise 6.4. Show that an operator Ψ is monotonic if and only if

$$\text{graph}\,\Psi(f) \subseteq \text{graph}\,\Psi(g) \text{ whenever } \text{graph} f \subseteq \text{graph} g$$

for every $f, g \in \mathscr{F}_1$.

Definition 6.6. An operator Ψ is said to be *finitary* or *continuous*[1] if, for every $x, y \in \mathbb{N}$ and $f \in \mathscr{F}_1$,

$$\Psi(f)(x) = y \quad \text{iff there exists finite } \vartheta \subseteq f \text{ such that } \Psi(\vartheta)(x) = y.$$

Recall that ϑ is a unary finite function if and only if $\vartheta \in f\mathscr{C}_1$ (see Problem 1.1).

Example 6.4. The operator

$$\Psi = \lambda f . (\lambda y . \mu x . f(x) = y)$$

is finitary. We must show that for every $f \in \mathscr{F}_1$ and $x, y \in \mathbb{N}$,

$$\Psi(f)(y) = x \quad \text{iff} \quad \text{there exists finite } \vartheta \subseteq f \text{ such that } \Psi(\vartheta)(y) = x.$$

Indeed:

($\rightarrow$) Assume that $\Psi(f)(y) = x$. Then,

$$f(x) = y, \quad \{0, \dots, x-1\} \subseteq \operatorname{dom} f \quad \text{and} \quad f(z) \neq y, \text{ for } z = 0, \dots, x-1.$$

We must find a finite function ϑ coinciding with f in the domain of ϑ. Since $0, \dots, x \in \operatorname{dom} f$, then $\operatorname{dom} \vartheta$ should be $\{0, \dots, x\}$. Therefore let

$$\vartheta = \lambda z . \begin{cases} f(z) & \text{if } z \leq x \\ \text{undefined} & \text{otherwise} \end{cases}.$$

Then:

(1) ϑ is a finite function;

(2) $\vartheta \subseteq f$ since $\operatorname{dom} \vartheta = \{0, \dots, x\} \subseteq \operatorname{dom} f$ and $\vartheta(z) = f(z)$ for every $z \leq 0$;

(3) $\Psi(\vartheta)(y) = x$. Observe that: $\vartheta(x) = f(x) = y$, $\{0, \dots, x-1\} \subseteq \operatorname{dom} \vartheta$ and $\vartheta(z) = f(z) \neq y$, for $z = 0, \dots, x-1$. Thus,

$$x = (\mu x . \vartheta(x) = y)$$

and so $\Psi(\vartheta)(y) = x$.

($\leftarrow$) Assume that there exists a finite $\vartheta \subseteq f$ such that $\Psi(\vartheta)(x) = y$. Then, $\vartheta(x) = y$, $\{0, \dots, x-1\} \subseteq \operatorname{dom} \vartheta$ and $\vartheta(z) \neq y$, for $z = 0, \dots, x-1$. Since $f(z) = \vartheta(z)$ for every $z \in \operatorname{dom} \vartheta$ then $f(x) = y$, $\{0, \dots, x-1\} \subseteq \operatorname{dom} f$ and $f(z) \neq y$, for $z = 0, \dots, x-1$. Hence,

$$x = (\mu x . f(x) = y)$$

and so $\Psi(f)(y) = x$.

---

[1]Finitary operators are continuous for the topology induced in $\mathscr{F}_1$ by the partial order $\subseteq$, hence the alternative name (see [32, 4]).

Exercise 6.5. Show that an operator is finitary if and only if for every $x, y \in \mathbb{N}$ and $f \in \mathscr{F}_1$,

$(x, y) \in \text{graph}\,\Psi(f)$ iff there exists finite $\vartheta \subseteq f$ such that $(x, y) \in \text{graph}\,\Psi(\vartheta)$,

that is,

$$\text{graph}\,\Psi(f) = \bigcup_{\substack{\vartheta \in f\mathscr{C}_1 \\ \vartheta \subseteq f}} \text{graph}\,\Psi(\vartheta).$$

6.3 Myhill-Shepherdson Theorem

We start by comparing finitary and monotonic operators.

Proposition 6.2. Every finitary operator is monotonic.

Proof. Let Ψ be finitary, $f \subseteq g$ and $\Psi(f)(x) = y$. We need to show that $\Psi(g)(x) = y$. Indeed, since Ψ is finitary, there is $\vartheta \in f\mathscr{C}_1$ such that:

$$\begin{cases} \vartheta \subseteq f \\ \Psi(\vartheta)(x) = y \end{cases}$$

and so

$$\begin{cases} \vartheta \subseteq g \\ \Psi(\vartheta)(x) = y \end{cases}.$$

Hence, again using the fact that Ψ is finitary, $\Psi(g)(x) = y$ as envisaged. □

Example 6.5. The operator

$$\Psi = \lambda f.(\lambda y. \mu x. f(x) = y)$$

in Example 6.4 is monotonic by Proposition 6.2.

We provide another possible application of Proposition 6.2.

Example 6.6. Consider the operator

$$\Delta = \lambda f. \lambda x. \begin{cases} 2f(x) & \text{if } x \in \text{dom}\, f \\ 2x + 1 & \text{otherwise} \end{cases}.$$

We start by showing that Δ is not monotonic. Hence, we must find $f, g \in \mathscr{F}_1$ such that

$$f \subseteq g \quad \text{but} \quad \Delta(f) \nsubseteq \Delta(g).$$

Let $f = \lambda x.\text{undefined}$ and $g = \lambda x.1$. Then $f \subseteq g$. Moreover,

$$\begin{cases} \Delta(f) = \lambda x.2x+1 \\ \Delta(g) = \lambda x.2 \end{cases}.$$

Thus, $\text{dom}\,\Delta(f) = \text{dom}\,\Delta(g) = \mathbb{N}$. However,

$$\Delta(f)(1) = 3 \neq 2 = \Delta(g)(1).$$

Hence $\Delta(f) \not\subseteq \Delta(g)$ and so Δ is not monotonic. Therefore, by Proposition 6.2, Δ is not finitary.

The converse of Proposition 6.2 does not hold in general as we now show.

Example 6.7. Consider the operator:

$$\Psi = \lambda f. \begin{cases} \lambda x.1 & \text{if } \text{dom}\,f = \mathbb{N} \\ \lambda x.\text{undefined} & \text{otherwise} \end{cases}$$

shown in Example 6.3 to be monotonic. We now show that Ψ is not finitary. For instance,

$$\Psi(id_{\mathbb{N}}) = \lambda x.1$$

while

$$\Psi(\vartheta) = \lambda x.\text{undefined}$$

for every $\vartheta \in f\mathscr{C}_1$. Hence,

$$\bigcup_{\substack{\vartheta \in f\mathscr{C}_1 \\ \vartheta \subseteq id_{\mathbb{N}}}} \text{graph}\,\Psi(\vartheta) = \emptyset \neq \mathbb{N} \times \{1\} = \text{graph}\,\Psi(id_{\mathbb{N}})$$

and so, by Exercise 6.5, Ψ is not finitary.

The following result shows that monotonicity together with an additional hypothesis entails finiteness.

Proposition 6.3. Assume that Ψ is a monotonic operator and that for every $f \in \mathscr{F}_1$ and $x \in \text{dom}\,\Psi(f)$ there is finite $\vartheta \subseteq f$ such that

$$\Psi(\vartheta)(x) = \Psi(f)(x).$$

Then, Ψ is finitary.

Proof. Consider arbitrary $x, y \in \mathbb{N}$ and $f \in \mathscr{F}_1$. We have to show:

($\rightarrow$) Assume that $\Psi(f)(x) = y$. So, $x \in \operatorname{dom} \Psi(f)$. Hence, by hypothesis, there is finite $\vartheta \subseteq f$ such that $\Psi(\vartheta)(x) = \Psi(f)(x)$ and, as a consequence, $\Psi(\vartheta)(x) = y$.

($\leftarrow$) Assume that there is a finite $\vartheta \subseteq f$ such that $\Psi(\vartheta)(x) = y$. Thus, $x \in \operatorname{dom} \Psi(\vartheta)$. Furthermore, since Ψ is assumed to be monotonic, $\Psi(\vartheta) \subseteq \Psi(f)$. So, $\Psi(f)(x) = \Psi(\vartheta)(x)$ and, therefore, $\Psi(f)(x) = y$. $\qquad\square$

Proposition 6.4. Every computable operator is finitary.

Proof. Let Ψ be a computable operator, P_Ψ a program that computes Ψ and $f \in \mathscr{F}_1$.

($\rightarrow$) Assume that $\Psi(f)(x) = y$. So, the execution, in the adopted computational model enriched with an oracle for f, of $\mathsf{P}_\Psi(\operatorname{Oracle}(f))(x)$ returns y in a finite number of steps and so $\operatorname{Oracle}(f)$ is invoked a finite number of times say on $z_1, \dots, z_n$ each of them in $\operatorname{dom} f$. Let

$$\vartheta = f|_{\{z_1, \dots, z_n\}} = \lambda z. \begin{cases} f(z) & \text{if } z \in \{z_1, \dots, z_n\} \\ \text{undefined} & \text{otherwise} \end{cases} : \mathbb{N} \rightharpoonup \mathbb{N}.$$

Observe that the execution, in the adopted computational model enriched with an oracle for f, of $\mathsf{P}_\Psi(\operatorname{Oracle}(f))(x)$ has the same value as the execution, in the adopted computational model enriched with an oracle for ϑ, of $\overline{\mathsf{P}_\Psi}(\operatorname{Oracle}(\vartheta))(x)$ where program $\overline{\mathsf{P}_\Psi}$ is obtained from P_Ψ by replacing all occurrences of $\operatorname{Oracle}(f)$ by $\operatorname{Oracle}(\vartheta)$. So $\Psi(\vartheta)(x) = \Psi(f)(x)$. Then, as required, $\vartheta \subseteq f$ and $\Psi(\vartheta)(x) = y$.

($\leftarrow$) Assume that there exists a finite $\vartheta \subseteq f$ such that $\Psi(\vartheta)(x) = y$. So, the execution, in the computational model enriched with an oracle for ϑ, of $\mathsf{P}_\Psi(\operatorname{Oracle}(\vartheta))(x)$ returns y in a finite number of steps and so $\operatorname{Oracle}(\vartheta)$ is invoked a finite number of times say on $z_1, \dots, z_n$ each of them in $\operatorname{dom} \vartheta$. Observe that the execution, in the adopted computational model enriched with an oracle for ϑ, of $\mathsf{P}_\Psi(\operatorname{Oracle}(\vartheta))(x)$ has the same value as the execution, in the adopted computational model enriched with an oracle for f, of $\overline{\mathsf{P}_\Psi}(\operatorname{Oracle}(f))(x)$ where $\overline{\mathsf{P}_\Psi}$ is obtained from P_Ψ by replacing all occurrences of $\operatorname{Oracle}(\vartheta)$ by $\operatorname{Oracle}(f)$. Thus, as envisaged, $\Psi(f)(x) = y$. $\qquad\square$

The converse does not hold in general. That is, there are finitary operators that are not computable operators.

Example 6.8. Let h be an enumeration of $\mathbb{K}^c$. Then, the operator

$$\Psi = \lambda f . h \circ f$$

is finitary but not computable. Indeed:

(1) Ψ is not a computable operator. Assume, by contradiction, that Ψ is a computable operator. Observe that $\Psi(id_\mathbb{N}) = h$. So, h would be a computable map, by Proposition 6.1, contradicting the fact that $\mathbb{K}^c$ is not a listable set, by Proposition 3.7.

(2) Ψ is a finitary operator. Indeed:

($\rightarrow$) Assume that $\Psi(f)(x) = y$. Take $\vartheta = f|_{\{x\}}$. Then,

$$\Psi(f|_{\{x\}})(x) = h(f|_{\{x\}}(x)) = h(f(x)) = \Psi(f)(x) = y.$$

($\leftarrow$) Assume that there is a finite function $\vartheta \subseteq f$ such that $\Psi(\vartheta)(x) = y$. Observe that

$$\Psi(f)(x) = h(f(x)) = h(\vartheta(x)) = \Psi(\vartheta)(x) = y$$

since $h(\vartheta(x)) = y$ entails that $x \in$ dom ϑ and, so, $f(x) = \vartheta(x)$.

Proposition 6.5. Every computable operator is monotonic.

Proof. Let Ψ be a computable operator. Then, by Proposition 6.4, Ψ is finitary. Hence, by Proposition 6.2, Ψ is monotonic. $\square$

After the necessary conditions proved above, the Myhill-Shepherdson Criterion Theorem below provides a necessary and sufficient condition for the computability of an operator. To this end, we need first the following lemma concerning equality of finitary operators.

Proposition 6.6. Let $\Psi, \Psi' : \mathscr{F}_1 \rightarrow \mathscr{F}_1$ be finitary and such that $\Psi|_{f\mathscr{C}_1} = \Psi'|_{f\mathscr{C}_1}$. Then, $\Psi = \Psi'$.

Proof. Let $f \in \mathscr{F}_1$ and $x, y \in \mathbb{N}$. Then,

$$
\begin{aligned}
\Psi(f)(x) = y \quad &\text{iff} \quad \text{there exists finite } \vartheta \subseteq f : \Psi(\vartheta)(x) = y \\
&\text{iff} \quad \text{there exists finite } \vartheta \subseteq f : \Psi'(\vartheta)(x) = y \quad \text{iff} \quad \Psi'(f)(x) = y.
\end{aligned}
$$

Therefore,

$$\Psi(f) = \Psi'(f)$$

for every $f \in \mathscr{F}_1$ and so $\Psi = \Psi'$. $\square$

The following theorem provides an alternative characterization of computable operator.

Proposition 6.7 (Myhill-Shepherdson Criterion Theorem). Let u be a proper universal function and $\Psi : \mathscr{F}_1 \rightarrow \mathscr{F}_1$. Then Ψ is computable if and only if Ψ is finitary and there exists a computable unary map s such that

$$\Psi(\phi_p^u) = \phi_{s(p)}^u$$

for every $p \in \mathbb{N}$.

Proof. ($\rightarrow$) Assume that Ψ is computable. So, by Proposition 6.4, Ψ is finitary. Furthermore, let

$$v = \lambda\, p,x.\, \Psi(\phi_p^u)(x) : \mathbb{N}^2 \rightharpoonup \mathbb{N}.$$

Since u and Ψ are computable, the following program computes v:

```
function (p,x) (
      return Pψ (function(z)(return Pu(p,z))) (x)
)
```

where P_Ψ and P_u are programs that compute Ψ and u, respectively. Moreover, since u is a proper universal function, there exists $s \in m\mathscr{C}_1$ such that

$$v(p,x) = u(s(p),x)$$

for every $p,x \in \mathbb{N}$. Therefore, for every $p,x \in \mathbb{N}$,

$$\Psi(\phi_p^u)(x) = v(p,x) = u(s(p),x) = \phi_{s(p)}^u(x)$$

and, so,

$$\Psi(\phi_p^u) = \phi_{s(p)}^u,$$

as required.

($\leftarrow$) Assume that Ψ is finitary and that we are given $s \in m\mathscr{C}_1$ such that

$$\Psi(\phi_p^u) = \phi_{s(p)}^u$$

for every $p \in \mathbb{N}$. Note that, using the fact that univ and u are isomorphic (thanks to Rogers' Theorem, Proposition 4.13), there exists a bijection $r \in m\mathscr{C}_1$, see Exercise 4.16, such that

$$\begin{cases} \phi_p^u = \phi_{r(p)}^{\text{univ}} \\ \phi_q^{\text{univ}} = \phi_{r^{-1}(q)}^u \end{cases}$$

for every $p,q \in \mathbb{N}$. Thus, in particular,

$$\Psi(\phi_q^{\text{univ}}) = \Psi(\phi_{r^{-1}(q)}^u) = \phi_{s(r^{-1}(q))}^u = \phi_{r(s(r^{-1}(q)))}^{\text{univ}}$$

for every $q \in \mathbb{N}$. Consider the operator Ψ' computed by the following program $P_{\Psi'}$:

```
function (o) (
      return function (x) (return Puniv(Pr(Ps(Pr−1(PJW(o)))),x))
)
```

where P_{univ}, P_r, P_s, $P_{r^{-1}}$ and P_{J_W} are programs that compute univ, r, s, r^{-1} and J_W, respectively. Furthermore, in particular for any finite unary ϑ and any program P_ϑ computing it, by Proposition 3.11,

$$\phi_{\widehat{P_\vartheta}}^{\text{univ}} = \vartheta.$$

Observe that $P_{\Psi'}(P_\vartheta)$ computes the function $\Psi'(\vartheta)$. Hence,

$$\Psi'(\vartheta) = P_{\Psi'}(P_\vartheta) = \text{univ}_{r(s(r^{-1}(\widehat{P_\vartheta})))} = \phi_{r(s(r^{-1}(\widehat{P_\vartheta})))}^{\text{univ}} = \Psi(\phi_{\widehat{P_\vartheta}}^{\text{univ}}) = \Psi(\vartheta).$$

Therefore, by Proposition 6.6, $\Psi = \Psi'$ and, so, Ψ is computable. $\qquad\square$

Observe that s lifts $\Psi|_{\mathscr{C}_1} : \mathscr{C}_1 \to \mathscr{C}_1$ to the level of indices. Hence, a finitary operator Ψ will be a computable operator if there is a computable way of finding the indices of $\Psi(f)$ from the indices of f, for every $f \in \mathscr{C}_1$.

Example 6.9. We show that there is $s \in \text{m}\mathscr{C}_1$ such that

$$\chi_{\text{dom}\,\phi_k}^{\text{p}} = \phi_{s(k)}$$

for every $k \in \mathbb{N}$. Consider the operator

$$\Psi = \lambda f . \chi_{\text{dom}\,f}^{\text{p}} : \mathscr{F}_1 \to \mathscr{F}_1.$$

Observe that

$$\Psi = \lambda f . (\lambda z . 1) \circ f.$$

Then, it is immediate to see that the following program

```
function (o) (
    return function (x) (return function(z)(return 1)(o(x)))
)
```

computes Ψ. Hence, by the Myhill-Shepherdson Criterion Theorem, there is $s \in \text{m}\mathscr{C}_1$ such that

$$\chi_{\text{dom}\,\phi_k}^{\text{p}} = \Psi(\phi_k) = \phi_{s(k)}$$

for every $k \in \mathbb{N}$, as required.

Definition 6.7. Let u be a universal function. A unary map s is said to be *u-extensional* if

$$\phi_{s(p_1)}^u = \phi_{s(p_2)}^u \quad \text{whenever} \quad \phi_{p_1}^u = \phi_{p_2}^u$$

for every $p_1, p_2 \in \mathbb{N}$.

Proposition 6.8. Let u be a universal function, $s \in \text{m}\mathscr{F}_1$ and $\Psi : \mathscr{F}_1 \to \mathscr{F}_1$. If

$$\Psi(\phi_p^u) = \phi_{s(p)}^u$$

for every $p \in \mathbb{N}$ then s is u-extensional.

Proof. Indeed, if $\phi_{p_1}^u = \phi_{p_2}^u$, then $\Psi(\phi_{p_1}^u) = \Psi(\phi_{p_2}^u)$ and, so,

$$\phi_{s(p_1)}^u = \Psi(\phi_{p_1}^u) = \Psi(\phi_{p_2}^u) = \phi_{s(p_2)}^u$$

as envisaged.   □

The reader should wonder here if, conversely, any given u-extensional map induces an operator in this way. The following variant of the Myhill-Shepherdson result provides the answer to this question for computable maps. Instead of providing a criterion for the computability of a given operator, this variant guarantees the existence, unicity and computability of a finitary operator specified by an extensional computable map.

Proposition 6.9 (Myhill-Shepherdson Existence Theorem). Let u be a proper universal function and s a computable u-extensional map. Then, there is a unique finitary operator $\Psi : \mathscr{F}_1 \to \mathscr{F}_1$ such that

$$\Psi(\phi_p^u) = \phi_{s(p)}^u$$

for every $p \in \mathbb{N}$. Furthermore, Ψ is computable.

Proof. First, we verify that there is a map $\Psi_0 : \mathscr{C}_1 \to \mathscr{C}_1$ such that

$$\Psi_0(\phi_p^u) = \phi_{s(p)}^u$$

for every $p \in \mathbb{N}$. Let $f \in \mathscr{C}_1$. Since u is universal, there is $p \in \mathbb{N}$ such that $\phi_p^u = f$. Take

$$\Psi_0(f) = \phi_{s(p)}^u.$$

Observe that in this way Ψ_0 is well defined precisely because s is assumed to be u-extensional. Indeed, if $f = \phi_{p_1}^u = \phi_{p_2}^u$ then $\Psi_0(f) = \phi_{s(p_1)}^u = \phi_{s(p_2)}^u$. Since u is proper universal, let $r : u \to \text{univ}$ be an isomorphim (thanks to Rogers' Theorem, Proposition 4.13). Consider the operator $\Psi : \mathscr{F}_1 \to \mathscr{F}_1$ computed by the following program P_Ψ:

```
function (o) (
    return function (x) (return P_univ(P_r(P_s(P_{r^{-1}}(P_{J_W}(o)))),x))
)
```

where P_{univ}, P_r, P_s, $\mathsf{P}_{r^{-1}}$ and P_{J_W} are programs that compute univ, r, s, r^{-1} and J_W, respectively. We now show that Ψ extends Ψ_0. Observe that, for each $f \in \mathscr{C}_1$ and any program P_f computing it, the function

$$\Psi(f)$$

is computed by $P_\Psi(P_f)$. Moreover, by Proposition 3.11,

$$\phi_{\widehat{P_f}}^{\text{univ}} = f$$

for every $f \in \mathscr{C}_1$. Thus,

$$
\begin{aligned}
\Psi(f) &= P_\Psi(P_f) \\
&= \text{univ}_{r(s(r^{-1}(\widehat{P_f})))} \\
&= \phi_{r(s(r^{-1}(\widehat{P_f})))}^{\text{univ}} \\
&= \phi_{r^{-1}(r(s(r^{-1}(\widehat{P_f}))))}^{u} \\
&= \phi_{s(r^{-1}(\widehat{P_f}))}^{u} \\
&= \Psi_0(\phi_{r^{-1}(\widehat{P_f})}^{u}) \\
&= \Psi_0(\phi_{r(r^{-1}(\widehat{P_f}))}^{\text{univ}}) \\
&= \Psi_0(\phi_{\widehat{P_f}}^{\text{univ}}) \\
&= \Psi_0(f)
\end{aligned}
$$

for each $f \in \mathscr{C}_1$, that is, $\Psi|_{\mathscr{C}_1} = \Psi_0$. Since Ψ is computable, then Ψ is finitary. Furthermore, Ψ is the unique finitary operator extending Ψ_0. Indeed, let Ψ' be a finitary operator such that $\Psi'|_{\mathscr{C}_1} = \Psi_0$. Then, $\Psi'|_{F\mathscr{C}_1} = \Psi|_{F\mathscr{C}_1}$ and, so, by Proposition 6.6, $\Psi' = \Psi$. $\qquad\square$

6.4 Kleene's Least Fixed Point Theorem

We now discuss fixed points of operators in the context of the partial order $\subseteq$ over $\mathscr{F}_1$ (see Exercise 4.5).

Definition 6.8. Let $\Psi : \mathscr{F}_1 \to \mathscr{F}_1$. A function $f \in \mathscr{F}_1$ is said to be a *fixed point* of Ψ if

$$\Psi(f) = f.$$

Furthermore, f is said to be a *least fixed point* of Ψ if:

- f is a fixed point of Ψ;

- $f \subseteq g$ for every fixed point g of Ψ.

Clearly, an operator has at most one least fixed point.

Example 6.10. Let

$$\Psi = \lambda f.\lambda x. \begin{cases} 1 & \text{if } x = 0 \\ xf(x-1) & \text{otherwise} \end{cases}.$$

We now show that $\lambda x.x!$ is a fixed point of Ψ. Indeed:

$$\Psi(\lambda x.x!) = \lambda x. \begin{cases} 1 & \text{if } x = 0 \\ x(\lambda x.x!)(x-1) & \text{otherwise} \end{cases}$$

$$= \lambda x. \begin{cases} 1 & \text{if } x = 0 \\ x(x-1)! & \text{otherwise} \end{cases}$$

$$= \lambda x.x!$$

Example 6.11. The operator

$$\Psi = \lambda f.\lambda x. \begin{cases} f(x) + 1 & \text{if } x \in \text{dom} f \\ 0 & \text{otherwise} \end{cases}$$

does not have fixed points. That is, there is no $f \in \mathscr{F}_1$ such that

$$\Psi(f) = f$$

as we now show. Indeed, assume, by contradiction that f is a fixed point of Ψ. Then,

$$f \subseteq \Psi(f) \text{ and } \Psi(f) \subseteq f.$$

Observe that $\text{dom} \Psi(f) = \mathbb{N}$. We consider two cases:

(i) $x \in \text{dom} f$. Then, $\Psi(f)(x) = f(x) + 1 \neq f(x)$ contradicting $f \subseteq \Psi(f)$.

(ii) $x \notin \text{dom} f$. Then, $\Psi(f)(x) = 0$ and $f(x)$ is undefined contradicting $\Psi(f) \subseteq f$.

Remark 6.3. Given an operator Ψ, we denote by

$$\text{lfp} \, \Psi$$

the (unique) least fixed point of Ψ whenever it exists.

Exercise 6.6. Given an operator Ψ with a least fixed point, show that if lfp Ψ is a map then it is the unique fixed point of Ψ.

We are ready to introduce a sufficient condition for an operator to have a least fixed point.

Proposition 6.10 (Kleene's Least Fixed Point Theorem). Every computable operator has a least fixed point. Moreover, the least fixed point is a computable function.

Proof. Let $\Psi : \mathscr{F}_1 \to \mathscr{F}_1$ be computable. Consider the family of functions $\{f_k\}_{k\in\mathbb{N}}$ defined as follows:

- $f_0 = \lambda x . \text{undefined}$;

- $f_{k+1} = \Psi(f_k)$.

We start by proving several properties of $\{f_k\}_{k\in\mathbb{N}}$:

(1) $f_k \in \mathscr{C}_1$ for every $k \in \mathbb{N}$. The proof follows by induction on k:

(Base) $f_0 = \lambda x . \text{undefined} \in \mathscr{C}_1$.

(Step) Assume, by induction hypothesis (IH), that $f_k \in \mathscr{C}_1$. Observe that $f_{k+1} = \Psi(f_k)$. So, $f_{k+1} \in \mathscr{C}_1$ by Proposition 6.1, since $f_k \in \mathscr{C}_1$ by IH and Ψ is computable.

(2) if $j \leq k$ then $f_j \subseteq f_k$ for every $j, k \in \mathbb{N}$. The proof is by induction on k as follows:

(Base) $k = 0$. Then $j = 0$ and $f_j \subseteq f_0$ because the relation $\subseteq$ is reflexive.

(Step) $k = k_0 + 1$. Assume, by induction hypothesis, that

$$\text{if } j \leq k_0 \text{ then } f_j \subseteq f_{k_0}$$

for every $j \in \mathbb{N}$. Then, since Ψ is monotonic, thanks to Exercise 6.5,

$$\text{if } j \leq k_0 \text{ then } \Psi(f_j) \subseteq \Psi(f_{k_0})$$

for every $j \in \mathbb{N}$, and, so,

$$\text{if } j \leq k_0 \text{ then } f_{j+1} \subseteq f_{k_0+1}$$

for every $j \in \mathbb{N}$. That is, replacing j by $i - 1$ and taking into account that $k_0 + 1 = k$,

$$\text{if } i - 1 \leq k - 1 \text{ then } f_i \subseteq f_k$$

for every $i \in \mathbb{N}^+$, and, so,

$$\text{if } i \leq k \text{ then } f_i \subseteq f_k$$

for every $i \in \mathbb{N}^+$. Moreover,

$$f_0 = \lambda x . \text{undefined} \subseteq f_k.$$

Therefore,

$$\text{if } j \leq k \text{ then } f_j \subseteq f_k$$

for every $j \in \mathbb{N}$.

(3) $\bigcup_{k\in\mathbb{N}} \text{graph } f_k$ is the graph of a unary function.

We have to check that

$$(x,y_1),(x,y_2) \in \bigcup_{k\in\mathbb{N}} \text{graph} f_k \quad \text{implies} \quad y_1 = y_2$$

holds for every $x,y_1,y_2 \in \mathbb{N}$. Indeed, assume that

$$(x,y_1),(x,y_2) \in \bigcup_{k\in\mathbb{N}} \text{graph} f_k.$$

Then, there are $k_1,k_2 \in \mathbb{N}$ such that

$$\begin{cases} (x,y_1) \in \text{graph} f_{k_1} \\ (x,y_2) \in \text{graph} f_{k_2} \end{cases}.$$

Without loss of generality assume that $k_1 \leq k_2$. Using (2), we conclude that

$$(x,y_1),(x,y_2) \in \text{graph} f_{k_2}$$

and, so, $y_1 = y_2$.

Capitalizing on the properties shown above, let f_∞ be the function such that

$$\text{graph} f_\infty = \bigcup_{k\in\mathbb{N}} \text{graph} f_k.$$

We now prove that f_∞ is computable and is the least fixed point of Ψ:

(1) graph f_∞ is listable. Since Ψ is computable, by the Myhill-Shepherdson Criterion Theorem (see Proposition 6.7), let $s \in m\mathscr{C}_1$ be such that

$$\Psi(\phi_p) = \phi_{s(p)}$$

for every $p \in \mathbb{N}$. Furthermore, let $q \in \text{ind}(f_0)$ and r be the unary map defined as follows:

- $r(0) = q$;

- $r(k+1) = s(r(k))$.

Clearly, map r is computable. In addition,

$$f_k = \phi_{r(k)}$$

as we now prove by induction on $k \in \mathbb{N}$. Indeed:

(Base) $k = 0$. Then,

$$f_0 = \phi_q = \phi_{r(0)}.$$

(Step) $k > 0$. Thus,

$$f_{k+1} = \Psi(f_k) = \Psi(\phi_{r(k)}) = \phi_{s(r(k))} = \phi_{r(k+1)}.$$

Therefore,

$$\bigcup_{k \in \mathbb{N}} \operatorname{graph} f_k = \bigcup_{k \in \mathbb{N}} \operatorname{graph} \phi_{r(k)}.$$

We now prove the listability of this set. Observe that

$$
\begin{aligned}
\bigcup_{k \in \mathbb{N}} \operatorname{graph} \phi_{r(k)} &= \bigcup_{k \in \mathbb{N}} \{(x,y) : \phi_{r(k)}(x) = y\} \\
&= \bigcup_{k \in \mathbb{N}} \{(x,y) : \operatorname{univ}(r(k),x) = y\} \\
&= \{(x,y) : \text{ such that } \operatorname{univ}(r(k),x) = y \text{ for some } k \in \mathbb{N}\} \\
&= \operatorname{proj}_{[2,3]}^{\mathbb{N}^3}(\{(k,x,y) : \operatorname{univ}(r(k),x) = y\}) \\
&= \operatorname{proj}_{[2,3]}^{\mathbb{N}^3}(\operatorname{graph}(\lambda\, k,x\,.\operatorname{univ}(r(k),x))).
\end{aligned}
$$

Let

$$g = \lambda\, k,x\,.\operatorname{univ}(r(k),x) = \operatorname{univ} \circ \langle r \circ \operatorname{proj}_{[1]}^{\mathbb{N}^2}, \operatorname{proj}_{[2]}^{\mathbb{N}^2}\rangle.$$

Then, it is immediate to see that g is computable. Hence, by the Graph Theorem (Proposition 2.13), the set

$$\operatorname{graph}(\lambda\, k,x\,.\operatorname{univ}(r(k),x)) = \operatorname{graph} g$$

is listable. Therefore, by Proposition 2.7,

$$\bigcup_{k \in \mathbb{N}} \operatorname{graph} \phi_{r(k)}$$

is listable because it is the image by the computable map $\operatorname{proj}_{[2,3]}^{\mathbb{N}^3}$ of the listable set $\operatorname{graph} g$. Thus, $\operatorname{graph} f_\infty$ is a listable set.

(2) f_∞ is computable, by (1) and the Graph Theorem (Proposition 2.13).

(3) $f_\infty \subseteq \Psi(f_\infty)$. We prove instead the equivalent statement:

$$\operatorname{graph} f_\infty \subseteq \operatorname{graph} \Psi(f_\infty).$$

Observe that

$$\operatorname{graph} f_j \subseteq \bigcup_{k \in \mathbb{N}} \operatorname{graph} f_k = \operatorname{graph} f_\infty$$

for every $j \in \mathbb{N}$. Hence, by the monotonicity of Ψ (by Proposition 6.5 since Ψ is computable) and Exercise 6.4, we have:

$$\text{graph}\,\Psi(f_j) \subseteq \text{graph}\,\Psi(f_\infty)$$

for every $j \in \mathbb{N}$. Hence,

$$\text{graph}\,f_{j+1} \subseteq \text{graph}\,\Psi(f_\infty)$$

for every $j \in \mathbb{N}$. That is,

$$\text{graph}\,f_i \subseteq \text{graph}\,\Psi(f_\infty)$$

for every $i \in \mathbb{N}^+$. Furthermore,

$$\text{graph}\,f_0 = \emptyset \subseteq \text{graph}\,\Psi(f_\infty).$$

Thus,

$$\text{graph}\,f_j \subseteq \text{graph}\,\Psi(f_\infty)$$

for every $j \in \mathbb{N}$, and, so,

$$\text{graph}\,f_\infty = \bigcup_{j \in \mathbb{N}} \text{graph}\,f_j \subseteq \text{graph}\,\Psi(f_\infty).$$

(4) $\Psi(f_\infty) \subseteq f_\infty$. We prove instead the equivalent statement:

$$\text{graph}\,\Psi(f_\infty) \subseteq \text{graph}\,f_\infty.$$

Let

$$(x,y) \in \text{graph}\,\Psi(f_\infty).$$

Then, since Ψ is finitary (thanks to Proposition 6.4 and Exercise 6.5 since Ψ is computable), there is finite $\vartheta \subseteq f_\infty$ such that

$$(x,y) \in \text{graph}\,\Psi(\vartheta).$$

Choose $j \in \mathbb{N}$ such that

$$\vartheta \subseteq f_j.$$

Note that such a j exists. Indeed, let

$$\text{graph}\,\vartheta = \{(x_1,y_1),\ldots,(x_n,y_n)\}.$$

Then, for each $i = 1,\ldots,n$ there is f_{k_i} such that

$$(x_i,y_i) \in \text{graph}\,f_{k_i}.$$

Thus, thanks to (2), in order to ensure that $\text{graph}\,\vartheta \subseteq \text{graph}\,f_j$ we take

$$j = \max\{k_1,\ldots,k_n\}.$$

Therefore, again by the finiteness of Ψ and Exercise 6.5, we have

$$(x,y) \in \text{graph}\,\Psi(f_j)$$

and, so,

$$(x,y) \in \text{graph}\,f_{j+1}.$$

Therefore,

$$(x,y) \in \bigcup_{k\in\mathbb{N}} \text{graph}\,f_k = \text{graph}\,f_\infty.$$

(5) $\Psi(f_\infty) = f_\infty$. Immediate from (3) and (4).

(6) Assume that $\Psi(g) = g$. Then, we prove that

$$f_j \subseteq g$$

by induction on $j \in \mathbb{N}$.

(Base) $j = 0$. Clearly, $f_0 = \lambda x.\text{undefined} \subseteq g$.

(Step) $j = j_0 + 1$. Assume, by the induction hypothesis, that

$$f_{j_0} \subseteq g.$$

Since Ψ is monotonic then:

$$f_j = f_{j_0+1} = \Psi(f_{j_0}) \subseteq \Psi(g) = g.$$

Therefore,

$$\text{graph}\,f_j \subseteq \text{graph}\,g$$

for every $j \in \mathbb{N}$. Thus

$$\text{graph}\,f_\infty = \bigcup_{j\in\mathbb{N}} \text{graph}\,f_j \subseteq \text{graph}\,g.$$

Hence, $f_\infty \subseteq g$. $\qquad\qquad\qquad\qquad\qquad\qquad\qquad\qquad\square$

Example 6.12. Consider the operator

$$\Psi = \lambda f.\lambda x.\begin{cases} 1 & \text{if } x = 0 \\ f(x+1) & \text{otherwise} \end{cases}.$$

We want to find the least fixed point of Ψ and to explain the general form of the fixed points of Ψ. The first step, towards applying Kleene's Least Fixed Point Theorem (Proposition 6.10), is to show that Ψ is a computable operator. Indeed, the program

```
function (f) (
    return function (x) (
        if x == 0 then
            return 1
        else
            return f(x+1)
    )
)
```

computes Ψ. Hence, there is lfp Ψ that we can compute using the proof of Kleene's Least Fixed Point Theorem. Observe that

1. $f_0 = \lambda x.\text{undefined}$

2.

$$f_1 = \Psi(f_0)$$

$$= \lambda x. \begin{cases} 1 & \text{if } x = 0 \\ f_0(x+1) & \text{otherwise} \end{cases}$$

$$= \lambda x. \begin{cases} 1 & \text{if } x = 0 \\ \text{undefined} & \text{otherwise} \end{cases}$$

3.

$$f_2 = \Psi(f_1)$$

$$= \lambda x. \begin{cases} 1 & \text{if } x = 0 \\ f_1(x+1) & \text{otherwise} \end{cases}$$

$$= \lambda x. \begin{cases} 1 & \text{if } x = 0 \\ \begin{cases} 1 & \text{if } x+1 = 0 \\ \text{undefined} & \text{otherwise} \end{cases} & \text{otherwise} \end{cases}$$

$$= \lambda x. \begin{cases} 1 & \text{if } x = 0 \\ \begin{cases} 1 & \text{if } x = -1 \\ \text{undefined} & \text{otherwise} \end{cases} & \text{otherwise} \end{cases}$$

$$= \lambda x. \begin{cases} 1 & \text{if } x = 0 \\ \text{undefined} & \text{otherwise} \end{cases}$$

4.

$$f_3 = \Psi(f_2)$$

$$= \lambda x. \begin{cases} 1 & \text{if } x = 0 \\ f_2(x+1) & \text{otherwise} \end{cases}$$

$$= \lambda x. \begin{cases} 1 & \text{if } x = 0 \\ \begin{cases} 1 & \text{if } x+1 = 0 \\ \text{undefined} & \text{otherwise} \end{cases} & \text{otherwise} \end{cases}$$

$$= \lambda x. \begin{cases} 1 & \text{if } x = 0 \\ \begin{cases} 1 & \text{if } x = -1 \\ \text{undefined} & \text{otherwise} \end{cases} & \text{otherwise} \end{cases}$$

$$= \lambda x. \begin{cases} 1 & \text{if } x = 0 \\ \text{undefined} & \text{otherwise} \end{cases}$$

...

We show that

$$f_k = \lambda x. \begin{cases} 1 & \text{if } x = 0 \\ \text{undefined} & \text{otherwise} \end{cases}$$

for every $k \in \mathbb{N}^+$ by induction on k.

(Base) Indeed,

$$f_1 = \lambda x. \begin{cases} 1 & \text{if } x = 0 \\ \text{undefined} & \text{otherwise} \end{cases}$$

(Step) In fact,

$$f_{k+1} = \Psi(f_k)$$

$$= \lambda x. \begin{cases} 1 & \text{if } x = 0 \\ f_k(x+1) & \text{otherwise} \end{cases}$$

$$= \lambda x. \begin{cases} 1 & \text{if } x = 0 \\ \begin{cases} 1 & \text{if } x+1 = 0 \\ \text{undefined} & \text{otherwise} \end{cases} & \text{otherwise} \end{cases}$$

$$= \lambda x. \begin{cases} 1 & \text{if } x = 0 \\ \begin{cases} 1 & \text{if } x = -1 \\ \text{undefined} & \text{otherwise} \end{cases} & \text{otherwise} \end{cases}$$

$$= \lambda x. \begin{cases} 1 & \text{if } x = 0 \\ \text{undefined} & \text{otherwise} \end{cases}$$

Therefore, by Kleene's Least Fixed Point Theorem,

$$\mathrm{lfp}\, \Psi = \lambda x. \begin{cases} 1 & \text{if } x = 0 \\ \text{undefined} & \text{otherwise} \end{cases}.$$

In what concerns other fixed points of Ψ, observe that $g \in \mathscr{F}_1$ is a fixed point of Ψ if it is of the form:

$$g = \Psi(g) = \lambda x. \begin{cases} 1 & \text{if } x = 0 \\ g(x+1) & \text{otherwise} \end{cases}.$$

That is, $g(x) = g(x+1)$ for every $x > 0$. Thus, for every $x > 0$, either $g(x)$ is undefined or $g(x) = k$ for some $k \in \mathbb{N}$. Therefore, all the fixed points of Ψ other than $\mathrm{lfp}\,\Psi$ are of the form:

$$\lambda x. \begin{cases} 1 & \text{if } x = 0 \\ k & \text{otherwise} \end{cases}$$

for every $k \in \mathbb{N}$. Thus Ψ has an infinite number of fixed points.

As a corollary of Kleene's Least Fixed Point Theorem (Proposition 6.10), using the Myhill-Shepherdson Existence Theorem (Proposition 6.9), we obtain a weaker formulation of the Recursion Theorem (Proposition 4.17).

Proposition 6.11. Let u be a proper universal function and $s \in \mathrm{m}\mathscr{C}_1$ an u-extensional map. Then, there exists $p \in \mathbb{N}$ such that:

$$\phi^u_{s(p)} = \phi^u_p.$$

Proof. Let $\Psi : \mathscr{F}_1 \to \mathscr{F}_1$ be a computable operator such that

$$\Psi(\phi^u_q) = \phi^u_{s(q)}$$

for every $p \in \mathbb{N}$. Such an operator exists, thanks to Myhill-Shepherdson Existence Theorem (Proposition 6.9), since u is proper universal and $s \in \mathrm{m}\mathscr{C}_1$ is u-extensional.

By Kleene's Least Fixed Point Theorem (Proposition 6.10), Ψ has a unique least fixed point which is computable. Choose

$$p \in \mathrm{ind}^u(\mathrm{lfp}\,\Psi).$$

Then, $\phi^u_{s(p)} = \Psi(\phi^u_p) = \Psi(\mathrm{lfp}\,\Psi) = \mathrm{lfp}\,\Psi = \phi^u_p$ and so p is such that $\phi^u_{s(p)} = \phi^u_p$. □

6.5 Solved Problems and Exercises

Problem 6.1. Show that

$$\Omega = \lambda f . \lambda z . \begin{cases} 1 & \text{if } \mathrm{dom}\, f = \mathbb{N} \text{ and range } f \subseteq \mathrm{ind}(f) \\ \text{undefined} & \text{otherwise} \end{cases}$$

is not a finitary operator.

Solution Observe that for every finite function ϑ we have

$$\mathrm{dom}\,\vartheta \neq \mathbb{N}$$

and so

$$\Omega(\vartheta) = \lambda z . \text{undefined}.$$

Hence, to show that Ω is not a finitary operator we must find a function f such that

$$\mathrm{dom}\,\Omega(f) \neq \emptyset.$$

Note that

$$\Omega(f) = \begin{cases} \lambda z . 1 & \text{if } \mathrm{dom}\, f = \mathbb{N} \text{ and range } f \subseteq \mathrm{ind}(f) \\ \lambda z . \text{undefined} & \text{otherwise} \end{cases}.$$

Therefore, for showing that Ω is not a finitary operator it is enough to find f such that

$$\mathrm{dom}\, f = \mathbb{N} \quad \text{and} \quad \text{range } f \subseteq \mathrm{ind}(f).$$

That is, for every $x \in \mathbb{N}$,

$$\mathrm{univ}_{f(x)} = f.$$

Let $v = \lambda q, x . q$. Then, by Problem 4.12, there is $p \in \mathbb{N}$ such that

$$\mathrm{univ}_p = v_p = \lambda x . p.$$

Take $f = \lambda x . p$. Then $\mathrm{dom}\, f = \mathbb{N}$ and range $f = \{p\} \subseteq \mathrm{ind}(f)$. ∇

Problem 6.2. Show that an operator Ψ is:

1. monotonic if and only if, for every $f, g \in \mathscr{F}_1$,

$$\text{graph}\,\Psi(f) \subseteq \text{graph}\,\Psi(g) \quad \text{whenever}\ \ \text{graph}\,f \subseteq \text{graph}\,g;$$

2. finitary if and only if, for every $f \in \mathscr{F}_1$,

$$\text{graph}\,\Psi(f) = \bigcup_{\substack{\vartheta \,\in\, f\mathscr{C}_1 \\ \vartheta\, \subseteq\, f}} \text{graph}\,\Psi(\vartheta).$$

Solution (1) We state without proving that for every $f, g \in \mathscr{C}_1$,

$$(\dagger) \quad f \subseteq g \ \text{ iff } \ \text{graph}\,f \subseteq \text{graph}\,g.$$

($\rightarrow$) Assume that (a) Ψ is monotonic and (b) $\text{graph}\,f \subseteq \text{graph}\,g$. From (b) and ($\dagger$), we can conclude that $f \subseteq g$ and so, by (a) we have $\Psi(f) \subseteq \Psi(g)$. Again, by ($\dagger$), we have $\text{graph}\,\Psi(f) \subseteq \text{graph}\,\Psi(g)$.

($\leftarrow$) Assume that (a) $\text{graph}\,f \subseteq \text{graph}\,g$ implies $\text{graph}\,\Psi(f) \subseteq \text{graph}\,\Psi(g)$ and that (b) $f \subseteq g$. By (b) and ($\dagger$), we can conclude that $\text{graph}\,f \subseteq \text{graph}\,g$ and so by (a) we have $\text{graph}\,\Psi(f) \subseteq \text{graph}\,\Psi(g)$. Again, by ($\dagger$), $\Psi(f) \subseteq \Psi(g)$.

(2)

($\rightarrow$) Assume that Ψ is a finitary operator. We must show that, for every $f \in \mathscr{F}_1$,

$$\text{graph}\,\Psi(f) = \bigcup_{\substack{\vartheta \,\in\, f\mathscr{C}_1 \\ \vartheta\, \subseteq\, f}} \text{graph}\,\Psi(\vartheta)$$

($\subseteq$) Suppose that $(x, y) \in \text{graph}\,\Psi(f)$. Then $\Psi(f)(x) = y$. Therefore, since Ψ is a finitary operator, there is $\vartheta \in f\mathscr{C}_1$ such that $\vartheta \subseteq f$ and $\Psi(\vartheta)(x) = y$. Thus, $(x, y) \in \text{graph}\,\Psi(\vartheta)$ and so

$$(x, y) \in \bigcup_{\substack{\vartheta \,\in\, f\mathscr{C}_1 \\ \vartheta\, \subseteq\, f}} \text{graph}\,\Psi(\vartheta).$$

($\supseteq$) We omit the proof since it is similar to the case ($\subseteq$).

($\leftarrow$) Let $f \in \mathscr{F}_1$ such that

$$(\ddagger) \quad \text{graph}\,\Psi(f) = \bigcup_{\substack{\vartheta \,\in\, f\mathscr{C}_1 \\ \vartheta\, \subseteq\, f}} \text{graph}\,\Psi(\vartheta)$$

Assume that $\Psi(f)(x) = y$ for some $x, y \in \mathbb{N}$. Then $(x, y) \in \operatorname{graph} \Psi(f)$ and so there is $\vartheta \in \mathsf{f}\mathscr{C}_1$ such that $\vartheta \subseteq f$ and $(x, y) \in \operatorname{graph} \Psi(\vartheta)$. Assume now that there is $\vartheta \in \mathsf{f}\mathscr{C}_1$ such that $\vartheta \subseteq f$ and $(x, y) \in \operatorname{graph} \Psi(\vartheta)$. then

$$(x, y) \in \bigcup_{\substack{\vartheta \in \mathsf{f}\mathscr{C}_1 \\ \vartheta \subseteq f}} \operatorname{graph} \Psi(\vartheta)$$

and so, by ($\ddagger$), $(x, y) \in \operatorname{graph} \Psi(f)$. $\qquad \nabla$

Problem 6.3. Show that the operator

$$\Psi = \lambda f . (\lambda y . (\mu x . f(x) = y)) : \mathscr{F}_1 \to \mathscr{F}_1$$

is computable.

Solution The given operator is computed by the following program P_Ψ:

```
function (o) (
    return function (y) (
        j = 0;
        while o(j) ≠ y do
            j = j + 1;
        return j
    )
)
```

Indeed, let $f \in \mathscr{F}_1$. We must show that $\mathsf{P}_\Psi(\operatorname{Oracle}(f))$ computes $\Psi(f)$ relatively to f. Observe that

$$\Psi(f)(y) = x \quad \text{iff} \quad 0, \ldots, x \in \operatorname{dom} f, f(j) \neq y \text{ for } j = 0, \ldots, x - 1 \text{ and } f(x) = y.$$

(1) Assume that $y \in \operatorname{dom} \Psi(f)$ and $\Psi(f)(y) = x$. Then, the execution relatively to f of $\operatorname{Oracle}(f)$ on input j for $j = 0, \ldots, x - 1$ terminates in a finite number of steps returning a value different from y. Moreover, the execution relatively to f of $\operatorname{Oracle}(f)$ on input x terminates in a finite number of steps returning y. Hence, the guard of the while will be true from 0 to $x - 1$ and the while terminates for x. So, the execution relatively to f of $\mathsf{P}_\Psi(\operatorname{Oracle}(f))$ on y terminates in a finite number of steps with x as the result.

(2) Assume that $y \notin \operatorname{dom} \Psi(f)$. Then, either (a) $f(j) \neq y$ for every $j \in \mathbb{N}$ or (b) there is $x \in \mathbb{N}$ such that $f(x) = y$ and $z \notin \operatorname{dom} f$ for some $0 \leq z \leq x$. Then the execution relatively to f of $\mathsf{P}_\Psi(\operatorname{Oracle}(f))(y)$ does not terminate. Indeed, if (a) holds then the guard of the while is always true and so j is always increasing. In case (b), the execution of $\operatorname{Oracle}(f)$ on z does not terminate. $\qquad \nabla$

Problem 6.4. Provide an example of a non-computable operator.

Solution Consider the operator

$$\Psi = \lambda f . \chi_{\mathbb{K}^c}^P.$$

By Proposition 6.1, every computable operator maps computable functions to computable functions. Hence, if Ψ were computable then $\chi_{\mathbb{K}^c}^P$ would also be computable because

$$\Psi(\lambda x . \text{undefined}) = \chi_{\mathbb{K}^c}^P$$

and λx.undefined is computable. But $\chi_{\mathbb{K}^c}^P$ is not computable thanks to Proposition 3.6. Hence, Ψ is not computable.   ▽

Problem 6.5. Find fixed points of the following operators:

1. $\Psi_1 = \lambda f . f$;

2. $\Psi_2 = \lambda f . \lambda x . \begin{cases} 0 & \text{if } x = 0 \\ 2x - 1 + f(x-1) & \text{otherwise} \end{cases}$.

Solution (1) Observe that Ψ_1 is clearly a computable operator. Therefore, Kleene's Least Fixed Point Theorem (Proposition 6.10) guarantees the existence of a least fixed point: λx.undefined. Observe that any function in $\mathscr{F}_1$ is a fixed point of Ψ_1.

(2) Since Ψ_2 is computable (see Exercise 6.12), we can use the proof of Kleene's Least Fixed Point Theorem (Proposition 6.10) to determine the least fixed point of Ψ_2. Observe that

$f_0 = \lambda x . \text{undefined}$

$f_1 = \Psi_2(f_0) = \lambda x . \begin{cases} 0 & \text{if } x = 0 \\ \text{undefined} & \text{otherwise} \end{cases}$

$f_2 = \Psi_2(f_1) = \lambda x . \begin{cases} 0 & \text{if } x = 0 \\ 1 & \text{if } x = 1 \\ \text{undefined} & \text{otherwise} \end{cases}$

$f_3 = \Psi_2(f_2) = \lambda x . \begin{cases} 0 & \text{if } x = 0 \\ 1 & \text{if } x = 1 \\ 4 & \text{if } x = 2 \\ \text{undefined} & \text{otherwise} \end{cases}$

$$f_4 = \Psi_2(f_3) = \lambda x. \begin{cases} 0 & \text{if } x = 0 \\ 1 & \text{if } x = 1 \\ 4 & \text{if } x = 2 \\ 9 & \text{if } x = 3 \\ \text{undefined} & \text{otherwise} \end{cases} .$$

We now show that:

$$f_k = \lambda x. \begin{cases} x^2 & \text{if } x < k \\ \text{undefined} & \text{otherwise} \end{cases}$$

by induction on k:

(Base) $k = 0$. Indeed,

$$\begin{aligned} f_0 &= \lambda x. \text{undefined} \\ &= \lambda x. \begin{cases} x^2 & \text{if } x < 0 \\ \text{undefined} & \text{otherwise} \end{cases} \end{aligned}$$

(Step) $k > 0$. In fact,

$$f_{k+1} = \Psi_2(f_k)$$

$$= \lambda x. \begin{cases} 0 & \text{if } x = 0 \\ 2x - 1 + f_k(x-1) & \text{otherwise} \end{cases}$$

$$= \lambda x. \begin{cases} 0 & \text{if } x = 0 \\ 2x - 1 + \begin{cases} (x-1)^2 & \text{if } x - 1 < k \\ \text{undefined} & \text{otherwise} \end{cases} & \text{otherwise} \end{cases}$$

$$= \lambda x. \begin{cases} 0 & \text{if } x = 0 \\ \begin{cases} 2x - 1 + x^2 - 2x + 1 & \text{if } x < k + 1 \\ \text{undefined} & \text{otherwise} \end{cases} & \text{otherwise} \end{cases}$$

$$= \lambda x. \begin{cases} 0 & \text{if } x = 0 \\ \begin{cases} x^2 & \text{if } x < k + 1 \\ \text{undefined} & \text{otherwise} \end{cases} & \text{otherwise} \end{cases}$$

$$= \lambda x. \begin{cases} x^2 & \text{if } x < k + 1 \\ \text{undefined} & \text{otherwise} \end{cases}$$

Hence, by Kleene's Least Fixed Point Theorem (Proposition 6.10), Ψ_2 has a least fixed point f_∞ such that the respective graph is the union of the graphs of the functions $f_0, f_1, \ldots$. That is,

$$f_\infty = \lambda x . x^2$$

is the least fixed point of Ψ. Observe that it is the unique fixed point of Ψ since f_∞ is a map and $f_\infty \subseteq f$ for every fixed point f of Ψ. ∇

Problem 6.6. Investigate the decidability and the listability of the set

$$P = \{p \in \mathbb{N} : \mathrm{lfp}\,\Psi \subseteq \phi_p\}$$

assuming that Ψ is a computable operator.

Solution Let

$$A = \{f \in \mathscr{C}_1 : \mathrm{lfp}\,\Psi \subseteq f\}.$$

Then, $\mathrm{ind}(A) = P$ since univ is a universal function. Observe that

$$A \neq \emptyset$$

since $\mathrm{lfp}(\Psi) \in A$ because Ψ has a computable least fixed point (Ψ is computable by hypothesis) by Kleene's Least Fixed Point Theorem (Proposition 6.10).

(Decidibility) By Rice's Theorem (Proposition 4.4), as $A \neq \emptyset$, P is a decidable set iff $A = \mathscr{C}_1$. Since $\lambda x.\,\mathrm{undefined} \in \mathscr{C}_1$ we can conclude that

$$P \text{ is decidable} \quad \text{iff} \quad \mathrm{lfp}\,\Psi = \lambda x.\,\mathrm{undefined}.$$

(Listability) We have two cases to consider:

(1) $\mathrm{lfp}\,\Psi$ is not a finite function. Then, P is not a listable set. Indeed, assume by contradiction, that P is a listable set. Then, by Rice-Shapiro's Theorem (Proposition 4.5), since $\mathrm{lfp}\,\Psi \in A$, there is a finite function $\vartheta \in A$ such that $\vartheta \subseteq \mathrm{lfp}(\Psi)$. Hence, $\mathrm{lfp}\,\Psi \subseteq \vartheta$ since $\vartheta \in A$ and so $\mathrm{lfp}\,\Psi = \vartheta$ contradicting the fact that $\mathrm{lfp}\,\Psi$ is a not a finite function.

(2) $\mathrm{lfp}\,\Psi$ is a finite function. Observe that

$$Q = \{q \in \mathbb{N} : \mathrm{lfp}\,\Psi \subseteq \theta_q\}$$

is a listable set. We now show that, for every $f \in \mathscr{C}_1$,

$$f \in A \quad \text{iff} \quad \text{there exists } q \in Q \text{ such that } \theta_q \subseteq f.$$

Indeed, let $f \in \mathscr{C}_1$:

($\rightarrow$) Suppose that $f \in A$. Then, $\mathrm{lfp}(\Psi) \subseteq f$. Let $q \in \mathbb{N}$ be such that

$$\mathrm{lfp}(\Psi) = \theta_q$$

since θ is a universal function for $f\mathscr{C}_1$ and, by hypothesis, $\mathrm{lfp}(\Psi) \in f\mathscr{C}_1$. Moreover, $q \in Q$ since $\mathrm{lfp}(\Psi) \subseteq \theta_q$.

($\leftarrow$) Assume that there is $q \in Q$ such that

$$\theta_q \subseteq f.$$

Then $\mathrm{lfp}(\Psi) \subseteq \theta_q$. Thus $\mathrm{lfp}(\Psi) \subseteq f$ and so $f \in A$.

So, by the Rice-Shapiro-McNaughton-Myhill Theorem (Proposition 4.5), $\mathrm{ind}(A) = P$ is a listable set. ∇

Supplementary Exercises

Exercise 6.7. Show that:

1. Operator $\lambda f . \chi^{P}_{\mathrm{dom}\,f}$ is computable.

2. Operator $\lambda f . \chi_{\mathrm{dom}\,f}$ is not computable.

Exercise 6.8. Analyze the computability of each of the following operators:

1. $\lambda f . \chi^{P}_{\mathrm{range}\,f}$;

2. $\lambda f . \chi_{\mathrm{range}\,f}$.

Exercise 6.9. Show that the operator $\lambda f . (\lambda z . \chi^{P}_{\mathrm{graph}\,f}(\beta_1^2(z)))$ is finitary using the definition.

Exercise 6.10. Classify the operator

$$\lambda f . \lambda x . \begin{cases} \text{undefined} & \text{if } \mathrm{dom}\,f \text{ is finite} \\ f(x) & \text{otherwise} \end{cases}.$$

Exercise 6.11. Consider the operator

$$\Psi = \lambda f . (\lambda x . \mathrm{pred}(f(x))) : \mathscr{F}_1 \to \mathscr{F}_1$$

where

$$\mathrm{pred} = \lambda x . \begin{cases} 0 & \text{if } x = 0 \\ x - 1 & \text{otherwise} \end{cases}.$$

Show that there is $s \in m\mathscr{C}_1$ such that

$$\Psi(\phi_k) = \phi_{s(k)}$$

using and not using the Myhill-Shepherdson Criterion Theorem.

Exercise 6.12. Consider the operator:

$$\Psi = \lambda f . \lambda x . \begin{cases} 0 & \text{if } x = 0 \\ 2x - 1 + f(x - 1) & \text{otherwise} \end{cases}.$$

1. Show that Ψ is finitary;

2. Find $s \in m\mathscr{C}_1$ such that $\Psi(\phi_p) = \phi_{s(p)}$ for every $p \in \mathbb{N}$;

3. What can be said about the computability of Ψ?

Exercise 6.13. Consider the operator

$$\Psi = \lambda g . \lambda x . \begin{cases} 0 & \text{if } x = 0 \\ 2 + g(x - 1) & \text{otherwise} \end{cases}.$$

What can be said about lfp Ψ?

Chapter 7

Turing Computational Model

In this book, we develop the theory of computability using an abstract high-level programming language for formalizing the notion of computable function. However, no student of computability theory can in the end avoid knowing some of the abstract computational models that have been widely used for the formalization of the notion of computable function. We briefly analyze Turing's approach to computability (for an advanced study of Turing concepts, the reader should consult [63, 25, 64, 8, 59]). An historical perspective of computability can be found in [50].

7.1 Turing Computable Functions

There are many versions of the Turing machine used in the literature, although all equivalent in the sense that they identify the same class of computable functions. Unless explicitly stated otherwise, herein the notion of Turing machine is as follows.

Definition 7.1. A (binary) *Turing machine* is a tuple

$$(A, \square, Q, q_s, q_h, \mathsf{L}, \mathsf{R}, \delta)$$

where:

- A is a finite set (the *alphabet*) containing the symbols 0 and 1;

- $\square \in A$ (the *blank space symbol*);

- Q is a finite set (the *state space*) such that $Q \cap A = \emptyset$;

- $q_s \in Q$ (the *starting state*);

- $q_h \in Q$ (the *halting state*) such that $q_h \neq q_s$;

- L (*left move instruction*);

- R (*right move instruction*);

- $\delta : Q \times A \rightharpoonup A \times Q \times \{L, R\}$ (the *transition function*) such that, for every $a \in A$, $(q_h, a) \notin \text{dom } \delta$.

As depicted in Figure 7.1, a Turing machine operates by reading and writing one symbol at a time on an infinite tape (infinite in both directions). To this end, the machine is also able to move along the tape in both directions, shifting one position at a time.

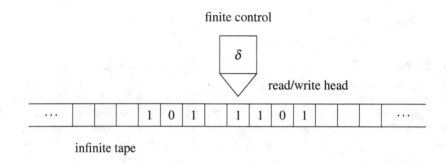

Figure 7.1: A Turing machine.

Given a Turing machine T,

$$\delta(q, a) = (q', a', M)$$

means that machine T, when in state q and reading a on the tape cell under the read/write head, writes a' on that tape cell, changes to state q' and moves the read/write head:

- to the left adjacent tape cell if M is L;

- to the right adjacent tape cell if M is R.

Example 7.1. Consider the machine T_0 defined as follows:

- $A = \{0, 1, \square\}$;

- $Q = \{q_s, q_1, q_2, q_h\}$;

- $\delta(q_s, a) = (a, q_1, R)$;

- $\delta(q_1,b) = (\square,q_1,\mathsf{R})$;

- $\delta(q_1,\square) = (\square,q_2,\mathsf{L})$;

- $\delta(q_2,a) = (0,q_h,\mathsf{L})$;

where $a \in \{0,1,\square\}$ and $b \in \{0,1\}$. This machine can be graphically described by the *transition diagram* as in Figure 7.2.

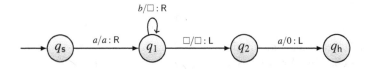

Figure 7.2: Transition diagram for Turing machine T_0.

In the sequel, in order to simplify the presentation, we describe Turing machines by their transition diagram.

We assume that the non-blank part of the tape is finite. Nevertheless, it is essential that the non-blank part can grow as needed (unbounded memory work space), hence the infinite tape. Since the two infinite blank parts are irrelevant and so is the placement of the non-blank part, it is more convenient to work with the following finitary notion.

Definition 7.2. By a *tape description* of a Turing machine T we mean a triple (u,a,v) represented by

u_1	$\cdots$	u_m	a	v_1	$\cdots$	v_n

where

- $a \in A$ (intended to be the *symbol being scanned* by the read/write head of T);

- $u \in A^*$ (intended to be the *finite sequence of symbols to the left* of the head until the left infinite blank part of the tape);

- $v \in A^*$ (intended to be the *finite sequence of symbols to the right* of the head until the right infinite blank part of the tape).

Observe that a tape description specifies not only a finite sequence of symbols but also distinguishes a position in that sequence (indicating which symbol is under the head). Moreover, the machine has no access to the "addresses" of the tape cells but only to what is written in each of them.

Definition 7.3. A *configuration* of a Turing machine T is a pair $(q,(u,a,v))$ represented by

$$uqav$$

or, graphically by

$$q: \boxed{u_1} \cdots \boxed{u_m} \boxed{a} \boxed{v_1} \cdots \boxed{v_n}$$

where $q \in Q$ and (u,a,v) is a tape description. Henceforth

$$\text{config } T$$

denotes the set of all configurations of T. A configuration $uqav$ is said to be *initial* if $q = q_s$. A configuration $uqav$ is said to be *terminal* if $q = q_h$.

Definition 7.4. Let T be a Turing machine. The transition relation

$$\rightarrow_T$$

over config T induced by δ is defined as follows:

- $ubqav \rightarrow_T uq'ba'v$ if $\delta(q,a) = (a',q',\mathsf{L})$;
- $uqabv \rightarrow_T ua'q'bv$ if $\delta(q,a) = (a',q',\mathsf{R})$.

Clearly, for every $c \in$ config T, there is at most one $c' \in$ config T such that $c \rightarrow_T c'$. Moreover, if c' exists then c is not terminal.

Definition 7.5. By a *computation segment* of a Turing machine T we understand a finite non-empty sequence $c_1 \ldots c_n$ of configurations of T such that

$$c_i \rightarrow_T c_{i+1}$$

for every $i < n$. Such a computation segment is said to be an *initial computation segment* if c_1 is initial. By a *terminating computation* of T we mean an initial computation segment $c_1 \ldots c_n$ such that c_n is terminal. By a *non-terminating computation* of T we mean an infinite sequence $\lambda i.c_i$ such that, for every $n \in \mathbb{N}^+$, the finite sequence $c_1 \ldots c_n$ is an initial computation segment.

By a *computation* of T we mean a sequence of configurations of T such that it is a terminating computation if finite and a non-terminating one otherwise.

Example 7.2. Consider the Turing machine T_0 introduced in Example 7.1. Then,

$$uq_s\square 1\square v \quad u\square q_1 1\square v \quad u\square\square q_1\square v \quad u\square q_2\square\square v \quad uq_h\square 0\square v$$

is an example of a computation.

When dealing with (binary) Turing machines, the following notation becomes handy.

Remark 7.1. Given $x \in \mathbb{N}$, we denote by

$$\tilde{x}$$

the binary representation of x.

Example 7.3. For instance, $\tilde{6}$ is the sequence 110 of length 3 and $\tilde{0}$ is the sequence 0 of length 1.

With the notion of computation by a Turing machine, we are finally ready to state precisely when a function in $\mathscr{F}$ is computed by such a machine.

Definition 7.6. Let $m \in \mathbb{N}^+$ and $n \in \mathbb{N}$. A function $f \in \mathscr{F}_n^m$ is said to be *Turing computable* or *T-computable* if there is a Turing machine T such that, for each $x_1, \ldots, x_n \in \mathbb{N}$ and $u, v \in A^*$, when starting from the configuration

$$u q_s \square \tilde{x}_1 \square \ldots \square \tilde{x}_n \square v$$

- the machine goes through a terminating computation ending at a configuration

$$u' q_h \square \tilde{y}_1 \square \ldots \square \tilde{y}_m \square v',$$

  for some $u', v' \in A^*$, if $(x_1, \ldots, x_n) \in \operatorname{dom} f$ with $f(x_1, \ldots, x_n) = (y_1, \ldots, y_m)$;

- the machine enters a non-terminating computation if $(x_1, \ldots, x_n) \notin \operatorname{dom} f$.

In this situation, we say that function f is *computed* by T.

Remark 7.2. We denote by

$$\mathscr{C}_T$$

the class of all T-computable functions.

Example 7.4. The map

$$\lambda x.0 : \mathbb{N} \to \mathbb{N}$$

is in $\mathscr{C}_T$. We now show that the Turing machine T_0 introduced in Example 7.1 computes this map. That is, T_0 when starting from the initial configuration

$$u q_s \square \tilde{x} \square v$$

goes through a terminating computation ending at the configuration

$$u' q_h \square 0 \square v'$$

for every $x \in \mathbb{N}$ (observe that $x \in \operatorname{dom} \lambda x.0$). In fact, T_0 while moving to the right, replaces each symbol of $\tilde{x}$ by $\square$ until it reaches $\square$. Then it moves to the left keeping $\square$. After that it replaces the preceding $\square$ by 0 and moves left halting.

The following concept is useful when putting together Turing machines.

Definition 7.7. Let $m \in \mathbb{N}^+$ and $n \in \mathbb{N}$. We say that $f \in \mathscr{F}_n^m$ is *computed while preserving the right context*, in short *rcp-computed*, by Turing machine T if, for each $x_1, \ldots, x_n \in \mathbb{N}$ and $u, v \in A^*$, when starting from the configuration

$$uq_s\square\tilde{x}_1\square\ldots\square\tilde{x}_n\square v$$

- the machine goes through a terminating computation ending at a configuration

$$u'q_h\square\tilde{y}_1\square\ldots\square\tilde{y}_m\square v$$

  for some $u' \in A^*$, if $(x_1, \ldots, x_n) \in \text{dom} f$ with $f(x_1, \ldots, x_n) = (y_1, \ldots, y_m)$;

- the machine enters a non-terminating computation if $(x_1, \ldots, x_n) \notin \text{dom} f$.

In this situation, we say that function f is rcp-computed by T.

Example 7.5. As an illustration of right context preserving computation, we now show that

$$\lambda x. \begin{cases} 1 & \text{if } x \text{ is odd} \\ 0 & \text{otherwise} \end{cases}$$

is rcp-computable. We start by observing that, given x,

$$x \text{ is odd} \qquad \text{iff} \qquad \tilde{x} \text{ ends in } 1.$$

Our goal is to define a Turing machine that starting with configuration

$$uq_s\square\tilde{x}\square v$$

ends with configuration

$$\begin{cases} u'q_h\square1\square v & \text{if } \tilde{x} \text{ ends in } 1 \\ u''q_h\square0\square v & \text{if } \tilde{x} \text{ ends in } 0 \end{cases}.$$

To this end, the Turing machine should output the last bit of $\tilde{x}$. We present two Turing machines that compute the given function. Therein, a denotes any symbol of the alphabet, and b denotes a bit.

(1) The first Turing machine, T_{odd1}, inspects one bit at a time, checking if it is the last. If it is then it leaves it and parks the head on the left. Otherwise, it erases the bit and proceeds to the next one. We consider the following states:

- q_s initial state;

- q_1 working state for the next bit;

- q_2 testing state to determine if the sequence has ended;

- q_3 erasing state, replacing each bit by the blank symbol, except for the last;

- q_4 success state;

- q_h halting state.

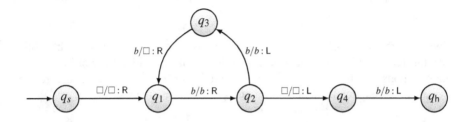

Figure 7.3: Transition diagram for Turing machine T_{odd1}.

The transition diagram is presented in Figure 7.3.

(2) The second Turing machine, T_{odd2}, takes into account that it is not necessary to erase the irrelevant bits. As we are interested only in the last bit, it is enough to move the head to the rightmost bit and then move back, keep the last bit and write a blank space on its left, leaving the rest on the input unchanged. The diagram for this second version can be found in Figure 7.4.

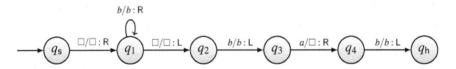

Figure 7.4: Transition diagram for Turing machine T_{odd2}.

Exercise 7.1. Define a Turing machine $T_{\lambda x_1, x_2 . (x_2, x_1)}$ that rcp-computes the function $\lambda x_1, x_2 . (x_2, x_1)$.

Example 7.6. Let $f_1, f_2 \in \mathscr{C}_T$ be unary functions. Then, let T_{f_1} and T_{f_2} be Turing machines that rcp-compute f_1 and f_2, respectively. We can define a Turing machine

$$T_{f_2 \circ f_1}$$

that rcp-computes $f_2 \circ f_1$, capitalizing on T_{f_1} and T_{f_2}. as follows:

- the set of states of $T_{f_2 \circ f_1}$ is the disjoint union of the sets of states of T_{f_1} and T_{f_2} but merging the halting state of T_{f_1} with the starting state of T_{f_1}, that we denote by q';

- the initial state q_s of $T_{f_2 \circ f_1}$ is the initial state of T_{f_1};

- the halting state q_h of $T_{f_2 \circ f_1}$ is the halting state of T_{f_2};

- the transition table δ of $_{f_2 \circ f_1}$ is obtained from the union of the transition tables δ_1 of T_{f_1} and δ_2 of T_{f_2}.

Note that there is no risk of ambiguous transitions. The only state were such ambiguous transitions could appear is q'. But, as this state is the halting state of T_{f_1} there are no transitions for this state in δ_1. The only transitions for q' in δ must come from δ_2. Hence, a computation from configuration $uq_s\square \tilde{x}\square v$ is as follows:

$$uq_s\square \tilde{x}\square v$$

$$\text{run } T_{f_1}$$

$$u'q'\square \widetilde{f_1(x)}\square v$$

$$\text{run } T_{f_2}$$

$$u''q_h\square \widetilde{f_2(f_1(x))}\square v.$$

Example 7.7. Let T_{f_1} and T_{f_2} be Turing machines that rcp-compute $f_1, f_2 : \mathbb{N} \to \mathbb{N}$ respectively, and $T_{\lambda x_1, x_2 . (x_2, x_1)}$ the Turing machine of Exercise 7.1. Then the Turing machine $T_{T_{f_1}|T_{f_2}}$ depicted in Figure 7.5 computes

Figure 7.5: Turing machine $T_{T_{f_1}|T_{f_2}}$.

$$\langle f_1, f_2 \rangle = \lambda x_1, x_2 . (f_1(x_1), f_2(x_2)),$$

that is, the aggregation of f_1 and f_2.

It should be clear by now to the reader that, within the Turing computational model, programming amounts to defining a Turing machine.

7.2 Turing Reducibility

As mentioned before, the notion of m-reducibility (Section 5.1), is quite restrictive in what we are allowed to use when building an algorithm for deciding problem C_1 from an algorithm deciding problem C_2. Indeed, we are allowed to ask for only one answer to problem C_2. An obvious generalization of m-reducibility would allow receiving answers to as many questions as needed on problem C_2. On the other hand, if C is decidable then one should expect $C \equiv_m C^c$. In fact, that is so, thanks to Proposition 5.1, provided that $\emptyset \subsetneq C \subsetneq \mathbb{N}$. This awkward role of sets $\emptyset$ and $\mathbb{N}$ stresses the need to look for other notions of reducibility. The notion of T-reducibility overcomes this issue but at the price of being less adequate concerning listability, since, as we shall see, $C \equiv_T C^c$ for every $C \in \mathscr{S}_1$.

To this end, we capitalize on the notion of relative computability presented in Section 6.1, to introduce the concepts of relative decidability and relative listability.

Definition 7.8. Let $C \subseteq \mathbb{N}$ and $f \in \mathscr{F}_1$. Then C is said to be *decidable relatively to f*, written

$$C \in \mathscr{D}_1/f$$

whenever $\chi_C \in \mathscr{C}_1/f$. Moreover, C is said to be *listable relatively to f*, written

$$C \in \mathscr{L}_1/f$$

whenever $\chi_C^p \in \mathscr{C}_1/f$.

Exercise 7.2. Show that $\mathbb{K}$ is decidable relatively to $\chi_{\mathbb{K}^c}^p$.

Exercise 7.3. State and prove criteria for listability relatively to some function f, adapting Proposition 2.11.

Returning to the issue of finding a more general form of reducibility, the following definition comes naturally, using relative computability.

Definition 7.9. Let $C_1, C_2 \in \mathscr{S}_1$. Then, C_1 is said to be *Turing reducible* or *T-reducible* to C_2, written

$$C_1 \leq_T C_2$$

whenever $\chi_{C_1} \in \mathscr{C}_1/\chi_{C_2}$. That is C_1 is decidable relatively to χ_{C_2}. Moreover, we say that C_1 and C_2 are *T-equivalent*, written

$$C_1 \equiv_T C_2$$

whenever $C_1 \leq_T C_2$ and $C_2 \leq_T C_1$.

Exercise 7.4. Show that if $C \leq_T D$ then $\mathscr{C}/\chi_C \subseteq \mathscr{C}/\chi_D$, and that if $C \equiv_T D$ then $\mathscr{C}/\chi_C = \mathscr{C}/\chi_D$.

Exercise 7.5. Show that $\leq_T$ is a preorder.

The following result shows that $\leq_T$ does generalize $\leq_m$ and that $\leq_T$ avoids the quirks of $\leq_m$. However, a price is paid in what regards listability.

Proposition 7.1. Let $C, D \in \mathscr{S}_1$. Then:

1. if $C \leq_m D$ then $C \leq_T D$;

2. $C \equiv_T C^c$;

3. if C is decidable then $C \leq_T D$;

4. if D is decidable and $C \leq_T D$ then C is decidable;

5. if C is listable then $C \leq_T \mathbb{K}$.

Proof. (1) Assume that $C \leq_m D$. Then, there is a computable map $s : \mathbb{N} \to \mathbb{N}$ such that $x \in C$ iff $s(x) \in D$ for every $x \in \mathbb{N}$. Then the following program

$$\text{function } (x) \ (\\ \quad \text{return } \text{Oracle}(\chi_D)(\mathsf{P}_s(x))\\ )$$

computes χ_C relatively to χ_D, where P_s is a program that computes s.
(2) We show that $C \leq_T C^c$. Observe that

$$\chi_C(x) = 1 - \chi_{C^c}(x).$$

Then, the following program

$$\text{function } (x) \ (\\ \quad \text{return } 1 - \text{Oracle}(\chi_{C^c})(x)\\ )$$

computes χ_C relatively to χ_{C^c}.
(3) Assume that C is a decidable set. Then χ_C is a computable map and so, by Exercise 6.3, $\chi_C \in \mathscr{C}_1 / \chi_D$, for every $D \subseteq \mathbb{N}$.
(4) Assume that D is a decidable set and $C \leq_T D$, Then, χ_C is computable relatively to χ_D. Thus, by Exercise 6.3, χ_C is also a computable map. Hence, C is a decidable set.
(5) Assume that C is a listable set. Then $C \leq_m \mathbb{K}$ by Proposition 5.3. Hence $C \leq_T \mathbb{K}$ by (1). $\qquad\qquad\square$

Exercise 7.6. Show that $\equiv_T$ is an equivalence relation.

7.3 Turing Degrees

We briefly introduce Turing reducibility. For more details the reader should consult for instance [8].

Definition 7.10. Given $C \in \mathscr{S}_1$,

$$d_T(C) = \{D \in \mathscr{S}_1 : C \equiv_T D\}$$

is the T-*degree* or *Turing degree* of C.

Exercise 7.7. Show the following assertions relating m-degrees with T-degrees:

1. For every $C \in \mathscr{S}_1$, $d_m(C) \subseteq d_T(C)$.

2. For every $C, D \in \mathscr{S}_1$, if $d_m(C) \leq_m d_m(D)$ then $d_T(C) \leq_T d_T(D)$.

Definition 7.11. By a *decidable* T-degree we mean a T-degree containing a decidable set, and by a *listable* T-degree we mean a T-degree containing a listable set.

Remark 7.3. We denote by

$$D_T \quad D_T^{\mathscr{D}} \quad D_T^{\mathscr{L}}$$

the class of all T-degrees, the class of all decidable ones and the class of all listable ones, respectively.

Exercise 7.8. Show that $\leq_T$ induces a partial order in D_T.

Remark 7.4. We denote by

$$\mathbf{0}_T \quad \text{and} \quad \mathbf{0}_T'$$

the T-degrees $d_T(\emptyset)$ and $d_T(\mathbb{K})$, respectively.

We now show some properties of T-degrees.

Proposition 7.2 (Structure of D_T)**.** The following assertions about T-degrees hold:

1. $\mathbf{0}_T \leq_T \mathbf{d}$ for every $\mathbf{d} \in D_T$;

2. $\mathbf{0}_T$ is the unique decidable T-degree;

3. If $\mathbf{d} \in D_T^{\mathscr{L}}$ then $\mathbf{d} \leq_T \mathbf{0}_T'$.

Proof. (1) Note that, by Proposition 7.1, $\emptyset \equiv_T \emptyset^c$ and so $\emptyset \leq_T \mathbb{N}$. Moreover, $\emptyset \leq_m D$ whenever $D \neq \mathbb{N}$ (Proposition 5.1) and so, by Proposition 7.1, $\emptyset \leq_T C$ for every set C.
(2) Let C be a decidable set. Then, $C \leq_T \emptyset$ by Proposition 7.1. Moreover, by (1), $\emptyset \leq_T C$.
(3) Let D be a listable set. Then, by Proposition 7.1, $D \leq_T \mathbb{K}$. □

The question of the existence of T-degrees between $\mathbf{0}_T$ and $\mathbf{0}_T'$ is answered positively by Dekker's Theorem [17] and by Friedberg-Muchnik Theorem [56].

7.4   Church-Turing Thesis

Many approaches have been proposed to define formally the notion of computable function, namely, among others: Turing machines (see Section 7.1), Kleene's recursive functions (see Definition 1.19), Church's λ-definable functions [9, 10, 33, 11, 4], Post's canonical systems [48, 49, 18, 45, 17], Shepherdson-Sturgis register machines [57, 17], Aanderaa-Cohen modular machines [1] and Diophantine sets [51, 41, 31]. The equivalence of the first three definitions in the list above inspired Stephen Kleene (in [34]) to state the following postulate, which has become known as the *Church-Turing thesis*:

> *Any function that can be accepted as computable is formally computable*
> *by some Turing machine.*

It has been shown that these and other formal notions of computability are all equivalent. In other words, any reasonable formalization of the notion of computability identifies only Turing computable functions. Notice that the Church-Turing thesis is not provable. Only on a case-by-case basis can be shown that a particular formalization of computability identifies as computable precisely the Turing computable functions.

The notion of computable function introduced herein using the abstract high-level programming language described in Section 1.2, was shown, in Proposition 1.9, to encompass all Kleene's recursive functions. On the other hand, using Gödelizations (see Section 2.4) and since $\lambda k.k : \mathbb{N} \to \mathbb{N}$ is a time constructible function (see [2]), a concept relevant for the implementation of the stceval function (see Section 1.2), the Turing computation model can accommodate our notion of computable function. Observe also that the stceval function was already defined in the URM computational model in [17].

The Church-Turing thesis is also used by some authors for justifying the computability of functions in an informal way and so avoiding a rigorous proof.

The discussion of the thesis in Kleene's book [35] is highly recommended (see also [60]). For a recent survey see [47].

7.5   Solved Problems and Exercises

Problem 7.1. Show that the map $\mathsf{S} = \lambda x.x + 1$ is in $\mathscr{C}_{\mathsf{T}}$.

Solution The Turing machine

$$T_{\mathsf{S}}$$

for computing S, for any $x \in \mathbb{N}$ and $u, v \in A^*$, when starting from configuration

$$u q_{\mathsf{s}} \square \tilde{x} \square v$$

should halt in a configuration

$$u'q_h\Box\widetilde{x+1}\Box v'.$$

Hence, the machine should move to the rightmost bit of the sequence $\tilde{x}$ and should add 1 to that first (least significant) bit. If that first bit is 0 then just replace it by 1 and move to the beginning of the sequence. If it is 1, then replace it by 0, move to the left bit and proceed as before, that is, until a 0 is found (and there is no bit to carry) or until the beginning of the sequence is reached. The transition diagram for T_S is presented in Figure 7.6. Therein, a denotes any symbol of the alphabet, and b denotes a bit.

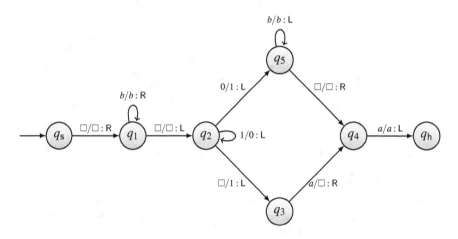

Figure 7.6: Transition diagram for Turing machine T_S.

For instance,

$$uaq_s\Box1\Box v \quad ua\Box q_11\Box v \quad ua\Box1q_1\Box v \quad ua\Box q_21\Box v \quad uaq_2\Box0\Box v$$

$$uq_3a10\Box v \quad u\Box q_410\Box v \quad uq_h\Box10\Box v$$

is an example of a computation of T_S. ∇

Problem 7.2. Define a Turing machine T_{23} that computes the constant 23.

Solution We need to define a machine that starting on an empty input

$$uq_s\Box\Box v$$

terminates with a final configuration

$$u'q_h\Box\widetilde{23}\Box v'.$$

Observe that

$$\tilde{23} \quad \text{is} \quad 10111.$$

Hence, the envisaged machine will sequentially write the string 10111, one symbol at a time, from left to right, and then park the head at the blank space on the left of the output. We present the transition diagram for T_{23} in Figure 7.7. Therein, a denotes any symbol of the alphabet, and b denotes a bit.

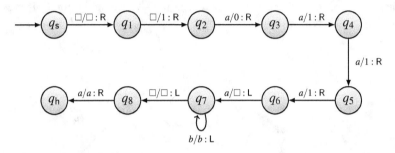

Figure 7.7: Transition diagram for T_{23}.

Observe that this machine does not meet the right context preserving requirement when computing 23. As $\tilde{23}$ is written from left to right, the right context is destroyed in the process. To define a machine meeting the right context preserving requirement it would be enough to write $\tilde{23}$ from right to left, not forgetting the blank on the left of the first 1.   ▽

Problem 7.3. Define a Turing machine $T_{\lambda x.(x,x)}$ that rcp-computes

$$\lambda x.(x,x).$$

Solution The underlying idea for this machine is not very complex. For each symbol in $\tilde{x}$ we will write the same symbol on the left part of the tape. To this end, at each step we need to know which symbol is being copied and to do so we extend the alphabet of the machine with two additional symbols: $*$ and $\#$. We will use $*$ to signal in the original word that a 0 is being copied and we will use $\#$ to signal that a 1 is being copied. So the alphabet A is $\{0,1,\square,\#,*\}$. The transition diagram for $T_{\lambda x.(x,x)}$ is depicted in Figure 7.8. Therein, a denotes any symbol of the alphabet, and b denotes a bit. We describe next its behaviour with an example. Consider the case $\tilde{5} = 101$. Initially, the machine is in the following configuration:

q_s :	x	x	x	x		1	0	1	

i.e., the machine is the initial state q_s and the head (the shaded gray cell) is parked on the left of the first symbol. Note that the cells to the left of the head might not be

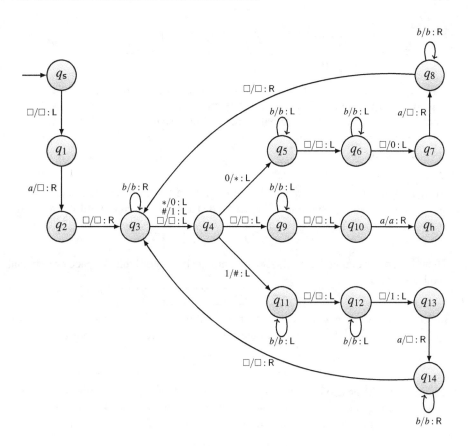

Figure 7.8: Transition diagram for Turing machine $T_{\lambda x.\,(x,x)}$.

necessarily empty (we write x to denote an arbitrary symbol). So, the machine creates an empty cell by writing a blank on the left:

q_2 : | x | x | x | | | 1 | 0 | 1 | |

The machine then scans the tape to the right until it finds a blank space or a special symbol * or #:

q_3 : | x | x | x | | | 1 | 0 | 1 | |

In this case, it finds a blank space (signalling the end of the word) so the machine moves left and analyzes the last symbol:

q_4 : | x | x | x | | | 1 | 0 | 1 |

The symbol 1 is found, which is replaced by #:

$$q_{11}: \quad \boxed{x} \ \boxed{x} \ \boxed{x} \ \boxed{\ } \ \boxed{\ } \ \boxed{1} \ \boxed{0} \ \boxed{\#}$$

Now, a copy of this 1 must be written to the left of the word and, afterward, the 1 must be restored in the original word. The machine moves left until it finds the first blank (the beginning of the original word) and then scans the additional symbols of the copy. In this first iteration there are none so it finds the blank cell that was created in the beginning:

$$q_{12}: \quad \boxed{x} \ \boxed{x} \ \boxed{x} \ \boxed{\ } \ \boxed{\ } \ \boxed{1} \ \boxed{0} \ \boxed{\#}$$

At this point, the machine writes the copy of 1 on the empty cell, creates another empty cell for the next symbol on the left and starts moving right until it reaches the end of the copy:

$$q_3: \quad \boxed{x} \ \boxed{x} \ \boxed{\ } \ \boxed{1} \ \boxed{\ } \ \boxed{1} \ \boxed{0} \ \boxed{\#}$$

Then, it starts scanning the original word until again it finds a blank space or a special symbol $*$ or #:

$$q_3: \quad \boxed{x} \ \boxed{x} \ \boxed{\ } \ \boxed{1} \ \boxed{\ } \ \boxed{1} \ \boxed{0} \ \boxed{\#}$$

In this case, it finds # which means that originally, there was 1 here, so it replaces it by 1 and moves left. To the right, all the symbols have already been copied:

$$q_4: \quad \boxed{x} \ \boxed{x} \ \boxed{\ } \ \boxed{1} \ \boxed{\ } \ \boxed{1} \ \boxed{0} \ \boxed{1}$$

Then it reads the next symbol and finds a 0. The machine replaces it by $*$ and now has to write it on the copy and then come back to restore the original 0. The sequence of (relevant) computation steps is the following:

$$q_5: \quad \boxed{x} \ \boxed{x} \ \boxed{\ } \ \boxed{1} \ \boxed{\ } \ \boxed{1} \ \boxed{*} \ \boxed{1}$$

$$q_6: \quad \boxed{x} \ \boxed{x} \ \boxed{\ } \ \boxed{1} \ \boxed{\ } \ \boxed{1} \ \boxed{*} \ \boxed{1}$$

$$q_3: \quad \boxed{x} \ \boxed{\ } \ \boxed{0} \ \boxed{1} \ \boxed{\ } \ \boxed{1} \ \boxed{*} \ \boxed{1}$$

$$q_3: \quad \boxed{x} \ \boxed{\ } \ \boxed{0} \ \boxed{1} \ \boxed{\ } \ \boxed{1} \ \boxed{*} \ \boxed{1}$$

$$q_4: \quad \boxed{x} \ \boxed{\ } \ \boxed{0} \ \boxed{1} \ \boxed{\ } \ \boxed{1} \ \boxed{0} \ \boxed{1}$$

$$q_{11}: \quad \boxed{x} \ \boxed{\ } \ \boxed{0} \ \boxed{1} \ \boxed{\ } \ \boxed{\#} \ \boxed{0} \ \boxed{1}$$

$$q_{12}: \quad \boxed{x} \ \boxed{\ } \ \boxed{0} \ \boxed{1} \ \boxed{\ } \ \boxed{\#} \ \boxed{0} \ \boxed{1}$$

$$q_3: \quad \boxed{\ } \ \boxed{1} \ \boxed{0} \ \boxed{1} \ \boxed{\ } \ \boxed{\#} \ \boxed{0} \ \boxed{1}$$

$$q_3: \quad \boxed{\ } \ \boxed{1} \ \boxed{0} \ \boxed{1} \ \boxed{\ } \ \boxed{\#} \ \boxed{0} \ \boxed{1}$$

$$q_4: \quad \boxed{\ } \ \boxed{1} \ \boxed{0} \ \boxed{1} \ \boxed{\ } \ \boxed{1} \ \boxed{0} \ \boxed{1}$$

q_9 :	1	0	1		1	0	1	

q_h :	1	0	1		1	0	1	

It is not very difficult to check that this machine meets the right context preserving requirement as it does not write beyond the last symbol of the original word. ▽

Problem 7.4. Show that, if D is a listable set relatively to χ_C, then

$$D \leq_T (\mathbb{K}/C),$$

where $(\mathbb{K}/C) = \{p \in \mathbb{N} : (p, p) \in \operatorname{dom} \operatorname{univ}^{\chi_C}\}$ and $\operatorname{univ}^{\chi_C}$ is the function computed relatively to χ_C by the program that computes univ.

Solution Assume that D is a listable set relatively to χ_C. Then,

$$\chi_D^P \in \mathscr{C}_1/\chi_C.$$

Thus, there is a program P that computes χ_D^P relatively to χ_C. For showing that $D \leq_T$ $(\mathbb{K}/C)$ we must prove that

$$\chi_D \in \mathscr{C}_1/\chi_{\mathbb{K}/C}.$$

Note that, for every $p \in \mathbb{N}$,

$$
\begin{aligned}
\chi_{\mathbb{K}/C}(p) = 1 \quad &\text{iff} \quad p \in \mathbb{K}/C \\
&\text{iff} \quad p \in \{p \in \mathbb{N} : (p, p) \in \operatorname{dom} \operatorname{univ}^{\chi_C}\} \\
&\text{iff} \quad (p, p) \in \operatorname{dom} \operatorname{univ}^{\chi_C} \\
&\text{iff} \quad \operatorname{univ}_p^{\chi_C}(p)\!\downarrow.
\end{aligned}
$$

Observe that if we manage to find $s \in m\mathscr{C}_1$ such that, for every $p \in \mathbb{N}$,

$$\operatorname{univ}_{s(p)}^{\chi_C} = \lambda x . \chi_D^P(p)$$

then,

$$
\begin{aligned}
\chi_{\mathbb{K}/C}(s(p)) = 1 \quad &\text{iff} \quad \operatorname{univ}_{s(p)}^{\chi_C}(s(p))\!\downarrow \\
&\text{iff} \quad (\lambda x . \chi_D^P(p))(s(p))\!\downarrow \\
&\text{iff} \quad \chi_D^P(p)\!\downarrow \\
&\text{iff} \quad p \in D \\
&\text{iff} \quad \chi_D(p) = 1.
\end{aligned}
$$

With the aim of using the *s-m-n* property of $\operatorname{univ}^{\chi_C}$ relatively to χ_C, consider the binary function v such that

$$v = \lambda p, x . \chi_D^P(p).$$

Using program P it is immediate to see that v is computable relatively to χ_C. Hence, there is $s \in m\mathscr{C}_1$ such that

$$\mathrm{univ}^{\chi_C}_{s(p)} = v_p = \lambda x . \chi^P_D(p)$$

for every $p \in \mathbb{N}$. Therefore,

$$\chi_D = \chi_{\mathbb{K}/C} \circ s.$$

Let P_s be a program for computing s. Thus the program

$$\mathsf{function}\ (x)\ (\mathsf{return}\ \mathsf{Oracle}(\chi_{\mathbb{K}/C})(P_s(x)))$$

computes χ_D relatively to $\chi_{\mathbb{K}/C}$. ∇

Supplementary Exercises

Exercise 7.9. Define a Turing machine T_F that when started in configuration

$$uq_s\square\tilde{x}_1\square\tilde{x}_2\square v$$

halts in configuration

$$u'q_h\square\mathrm{First}(\tilde{x}_2)\square\tilde{x}_1\square v'$$

where by $\mathrm{First}(\tilde{x}_2)$ we mean the first bit in the binary representation of x_2.

Exercise 7.10. Show that the function $\lambda x_1,x_2,x_3 . x_2$ is Turing computable.

Exercise 7.11. Define a Turing machine for rcp-computing each of the following functions:

1. $\lambda x . \begin{cases} 1 & \text{if } x > 0 \\ 0 & \text{otherwise} \end{cases}$;

2. $\lambda x_1,x_2 . \begin{cases} 1 & \text{if } x_1 \le x_2 \\ 0 & \text{otherwise} \end{cases}$;

3. $\lambda x . \begin{cases} x-1 & \text{if } x > 0 \\ \text{undefined} & \text{otherwise} \end{cases}$.

Exercise 7.12. Given Turing machines T_1 and T_2 that rcp-compute the unary functions f_1 and f_2, respectively, build a Turing machine that rcp-computes

$$\langle f_1, f_2 \rangle = \lambda x . (f_1(x), f_2(x)).$$

Exercise 7.13. Show that $\mathbb{K}/C$ is a listable set relatively to χ_C.

Exercise 7.14. Show that if A is a listable set relatively to χ_B and $B \le_T C$ then A is a listable set relatively to χ_C.

Part II

A Computability View of
Selected Topics

Chapter 8

Undecidability of Arithmetic

The main goals of this chapter are the proof of the semidecidability of the decision problem

> Given a formula φ and a set of formulas Γ, is φ derivable from Γ?

and the proof of the undecidability of theories of arithmetic (Gödel's First Incompleteness Theorem).

We start by presenting some essential notions of first-order logic namely language, Hilbert calculus and first-order theory.

8.1 First-Order Logic

The main components of a logic are the language or the set of formulas, a calculus for reasoning about formulas and a semantics for providing the meaning of formulas and sets of formulas. For the purpose of the chapter it is not needed to discuss the semantics of first-order logic.

Language

The set of first-order formulas is defined over a particular signature. Moreover, we assume a fixed decidable set X of variables.

Definition 8.1. A *first-order signature* (or, simply, *signature*) is a triple

$$\Sigma = (F, P, \tau)$$

such that F and P are decidable sets disjoint from each other and from X and $\tau :$ $F \cup P \to \mathbb{N}$ is a computable map. The elements of F are said to be *function symbols*.

The elements of P are said to be *predicate symbols*. Map τ returns the *arity* of its argument.

Remark 8.1. Given a first-order signature Σ and $n \in \mathbb{N}$, let:

$$F_n$$

denote the set $\{f \in F : \tau(f) = n\}$ of *function symbols of arity n* and

$$P_n$$

denote the set $\{p \in P : \tau(p) = n\}$ of *predicate symbols of arity n*. The elements of F_0 are also called *constant symbols*.

As usual, it is assumed that $P_0 = \emptyset$ and that $P \neq \emptyset$.

Exercise 8.1. Show that F_n and P_n are decidable for each $n \in \mathbb{N}$.

Example 8.1. Let $\Sigma_{\mathbb{N}}$ be the signature for the natural numbers defined as follows: $F_0 = \{0\}$, $F_1 = \{S\}$, $F_2 = \{+, \times\}$, $F_n = \emptyset$ for every $n > 2$, $P_2 = \{\cong, <\}$ and $P_n = \emptyset$ for $n \neq 2$.

Example 8.2. Let Σ_{RCOF} be the signature for real closed ordered fields defined as follows: $F_0 = \{0, 1\}$, $F_1 = \{-\}$, $F_2 = \{+, \times\}$, $F_n = \emptyset$ for every $n > 2$, $P_2 = \{\cong, <\}$ and $P_n = \emptyset$ for $n \neq 2$.

For first-order logic we should start by defining the concept of term.

Definition 8.2. The set T_Σ of *terms* over a signature Σ is inductively defined as follows:

- $X \cup F_0 \subseteq T_\Sigma$;

- $f(t_1, \ldots, t_n) \in T_\Sigma$ whenever $f \in F_n$, $t_1, \ldots, t_n \in T_\Sigma$ and $n \in \mathbb{N}^+$.

Exercise 8.2. Show that T_Σ is decidable.

Example 8.3. For instance, $+(S(x), S(S(0))) \in T_{\Sigma_{\mathbb{N}}}$ and $+(1, \times(-(x), 0)) \in T_{\Sigma_{\text{RCOF}}}$.

In order to simplify the presentation it is usual to use infix notation instead of prefix notation when presenting terms.

Example 8.4. Thus, we may write $S(x) + S(S(0))$ and $1 + ((-x) \times 0)$ instead of $+(S(x), S(S(0)))$ and $+(1, \times(-(x), 0))$, respectively.

In the context of a formula, it is important to distinguish between variables that occur free and variables that are bound to a quantifier. For this purpose, it is convenient to define first the set of variables that occur in a term.

Definition 8.3. The map $\text{var}_\Sigma : T_\Sigma \to \wp X$ assigning to each term the set of variables occurring in it is inductively defined as follows:

- $\text{var}_\Sigma(x) = \{x\}$;
- $\text{var}_\Sigma(c) = \emptyset$ for $c \in F_0$;
- $\text{var}_\Sigma(f(t_1,\ldots,t_n)) = \text{var}_\Sigma(t_1) \cup \cdots \cup \text{var}_\Sigma(t_n)$.

Example 8.5. For instance, $\text{var}_{\Sigma_\mathbb{N}}(1 + ((-x) \times 0)) = \{x\}$.

Exercise 8.3. Show that var_Σ is a computable map.

We are ready to define the first-order language over signature Σ.

Definition 8.4. The set L_Σ of *formulas* over Σ, called the *first-order language* over Σ, is inductively defined as follows:

- $p(t_1,\ldots,t_n) \in L_\Sigma$ whenever $p \in P_n$, $t_1,\ldots,t_n \in T_\Sigma$ and $n \in \mathbb{N}^+$;
- $(\neg\varphi) \in L_\Sigma$ whenever $\varphi \in L_\Sigma$;
- $(\varphi \supset \psi) \in L_\Sigma$ whenever $\varphi, \psi \in L_\Sigma$;
- $(\forall x\,\varphi) \in L_\Sigma$ whenever $x \in X$ and $\varphi \in L_\Sigma$.

Example 8.6. For instance, $(\forall x\,((S(x) + S(S(0))) \cong S(0))) \in L_{\Sigma_\mathbb{N}}$.

From now on, when no confusion arises, we may write formulas without some of the parentheses. For instance, we may write $\forall x\,S(x) + S(S(0)) \cong S(0)$.

We consider the usual abbreviations $\varphi_1 \vee \varphi_2$, $\varphi_1 \wedge \varphi_2$, $\varphi_1 \equiv \varphi_2$ and $\exists x\,\varphi$ for disjunction, conjunction, equivalence and existential quantification, respectively.

Exercise 8.4. Show that L_Σ is decidable.

Definition 8.5. The map $\text{fv}_\Sigma : L_\Sigma \to \wp X$ assigning to each formula the set of variables occurring *free* in the formula is inductively defined as follows:

- $\text{fv}_\Sigma(p(t_1,\ldots,t_n)) = \text{var}_\Sigma(t_1) \cup \cdots \cup \text{var}_\Sigma(t_n)$;
- $\text{fv}_\Sigma(\neg\varphi) = \text{fv}_\Sigma(\varphi)$;
- $\text{fv}_\Sigma(\varphi \supset \psi) = \text{fv}_\Sigma(\varphi) \cup \text{fv}_\Sigma(\psi)$;
- $\text{fv}_\Sigma(\forall x\,\varphi) = \text{fv}_\Sigma(\varphi) \setminus \{x\}$.

Moreover, we say that φ is a *closed formula* whenever $\text{fv}_\Sigma(\varphi) = \emptyset$.

Example 8.7. For instance, $\text{fv}_{\Sigma_\mathbb{N}}(\forall x\,S(x) \cong y) = \{y\}$.

Exercise 8.5. Show that fv_Σ is a computable map.

Now we introduce the notion of substitution of a variable by a term in a formula. We must start by defining substitution of variables by terms in terms.

Exercise 8.6. Let $y_1, \ldots, y_m$ be distinct variables and $u_1, \ldots, u_m$ be terms. Define inductively the map

$$\lambda t \,.\, [t]^{y_1,\ldots,y_m}_{u_1,\ldots,u_m} : T_\Sigma \to T_\Sigma$$

assigning to each term t, the term obtained by simultaneously and uniformly replacing each variable y_i in t by term u_i, for each $1 \leq i \leq m$.

Definition 8.6. Let $y_1, \ldots, y_m$ be distinct variables and $u_1, \ldots, u_m$ be terms. The map

$$\lambda \varphi \,.\, [\varphi]^{y_1,\ldots,y_m}_{u_1,\ldots,u_m} : L_\Sigma \to L_\Sigma$$

is inductively defined as follows:

- $[p(t_1,\ldots,t_n)]^{y_1,\ldots,y_m}_{u_1,\ldots,u_m}$ is $p([t_1]^{y_1,\ldots,y_m}_{u_1,\ldots,u_m}, \ldots, [t_n]^{y_1,\ldots,y_m}_{u_1,\ldots,u_m})$;

- $[\neg\, \varphi]^{y_1,\ldots,y_m}_{u_1,\ldots,u_m}$ is $\neg[\varphi]^{y_1,\ldots,y_m}_{u_1,\ldots,u_m}$;

- $[\varphi \supset \psi]^{y_1,\ldots,y_m}_{u_1,\ldots,u_m}$ is $[\varphi]^{y_1,\ldots,y_m}_{u_1,\ldots,u_m} \supset [\psi]^{y_1,\ldots,y_m}_{u_1,\ldots,u_m}$;

- $[\forall x\, \varphi]^{y_1,\ldots,y_m}_{u_1,\ldots,u_m}$ is:

    - $\forall x\, [\varphi]^{y_1,\ldots,y_m}_{u_1,\ldots,u_m}$ whenever $x \notin \{y_1, \ldots, y_m\}$;

    - $\forall x\, [\varphi]^{y_1,\ldots,y_m}_{u'_1,\ldots,u'_m}$ where u'_j is u_j for $j \neq k$ and u'_k is x whenever x is y_k.

Example 8.8. Observe that $[\forall x\, S(x) \cong y]^x_0$ is $\forall x\, S(x) \cong y$ and $[\forall x\, S(x) \cong y]^y_0$ is $\forall x\, S(x) \cong 0$.

Exercise 8.7. Show that $\lambda \varphi \,.\, [\varphi]^{y_1,\ldots,y_m}_{u_1,\ldots,u_m}$ is a computable map.

The following notion of free term for a variable in a formula helps in avoiding unwanted substitutions.

Definition 8.7. The set $\rhd_\Sigma \subseteq T_\Sigma \times X \times L_\Sigma$ is inductively defined as follows:

- $(t, x, p(t_1, \ldots, t_n)) \in \rhd_\Sigma$;

- $(t, x, \neg\, \varphi) \in \rhd_\Sigma$ whenever $(t, x, \varphi) \in \rhd_\Sigma$;

- $(t, x, \varphi \supset \psi) \in \rhd_\Sigma$ whenever $(t, x, \varphi) \in \rhd_\Sigma$ and $(t, x, \psi) \in \rhd_\Sigma$;

- $(t, x, \forall y\, \varphi) \in \rhd_\Sigma$ whenever

    (1) either y is x

(2) or the two following conditions are fulfilled:

$$\begin{cases} \text{if } x \in \text{fv}_\Sigma(\varphi) \text{ then } y \notin \text{var}_\Sigma(t); \\ (t,x,\varphi) \in \triangleright_\Sigma. \end{cases}$$

When $(t,x,\varphi) \in \triangleright_\Sigma$ it is said that *term t is free for variable x in formula φ*, usually written $t \triangleright_\Sigma x : \varphi$.

Example 8.9. Observe that $S(x) \triangleright_{\Sigma_\mathbb{N}} x : \forall x\, S(x) \cong y$ and $S(x) \not\triangleright_{\Sigma_\mathbb{N}} y : \forall x\, S(x) \cong y$.

Exercise 8.8. Show that $\triangleright_\Sigma$ is decidable.

Hilbert Calculus

A calculus for a logic includes a set of axioms and a set of inference rules. There are several ways to present a calculus. Herein, we define an Hilbert calculus for first-order logic. In such a calculus all the reasoning is done at the level of the formula. An *axiom* is a formula that is assumed to be always true. An inference rule is a pair composed by a finite set of formulas (the premises of the rule) and a formula (the conclusion of the rule).

Definition 8.8. Let $\Gamma \subseteq L_\Sigma$. The set $\Gamma^{+\Sigma}$ of formulas *derivable* from Γ is defined inductively as follows:

* $\Gamma \subseteq \Gamma^{+\Sigma}$ (*hypotheses*)

* $\Gamma^{+\Sigma}$ contains the *axioms*:

(Ax1) $\varphi \supset (\psi \supset \varphi) \in \Gamma^{+\Sigma}$

(Ax2) $(\varphi \supset (\psi \supset \delta)) \supset ((\varphi \supset \psi) \supset (\varphi \supset \delta)) \in \Gamma^{+\Sigma}$

(Ax3) $((\neg \varphi) \supset (\neg \psi)) \supset (\psi \supset \varphi) \in \Gamma^{+\Sigma}$

(Ax4) $(\forall x\, \varphi) \supset [\varphi]_t^x \in \Gamma^{+\Sigma}$ whenever $t \triangleright_\Sigma x : \varphi$

(Ax5) $(\forall x\, (\varphi \supset \psi)) \supset (\varphi \supset (\forall x\, \psi)) \in \Gamma^{+\Sigma}$ whenever $x \notin \text{fv}_\Sigma(\varphi)$

for each $\varphi, \psi, \delta \in L_\Sigma, x \in X$ and $t \in T_\Sigma$

* $\Gamma^{+\Sigma}$ is closed under the *rules*:

Modus Ponens (MP) $\psi \in \Gamma^{+\Sigma}$ whenever $\varphi, \varphi \supset \psi \in \Gamma^{+\Sigma}$

Generalization (Gen) $\forall x\, \varphi \in \Gamma^{+\Sigma}$ whenever $\varphi \in \Gamma^{+\Sigma}$

for each $\varphi, \psi \in L_\Sigma$ and $x \in X$.

Remark 8.2. Given a signature Σ, we denote by

$$\text{Ax}_\Sigma$$

the set of all axioms (see Definition 8.8) over Σ. Moreover, we may write

$$\vdash_\Sigma \varphi$$

instead of $\varphi \in \emptyset^{\vdash_\Sigma}$ for stating that φ is a *theorem*. Finally, we may write

$$\Gamma \vdash_\Sigma \varphi$$

for $\varphi \in \Gamma^{\vdash_\Sigma}$.

Example 8.10. We show that $\forall x \forall y\, \varphi \vdash_\Sigma \forall y \forall x\, \varphi$. Indeed,

1	$\forall x \forall y\, \varphi \in \{\forall x \forall y\, \varphi\}^{\vdash_\Sigma}$	Hyp
2	$(\forall x \forall y\, \varphi) \supset (\forall y\, \varphi) \in \{\forall x \forall y\, \varphi\}^{\vdash_\Sigma}$	Ax4
3	$\forall y\, \varphi \in \{\forall x \forall y\, \varphi\}^{\vdash_\Sigma}$	MP : 1,2
4	$(\forall y\, \varphi) \supset \varphi \in \{\forall x \forall y\, \varphi\}^{\vdash_\Sigma}$	Ax4
5	$\varphi \in \{\forall x \forall y\, \varphi\}^{\vdash_\Sigma}$	MP : 3,4
6	$\forall x\, \varphi \in \{\forall x \forall y\, \varphi\}^{\vdash_\Sigma}$	Gen : 5
7	$\forall y \forall x\, \varphi \in \{\forall x \forall y\, \varphi\}^{\vdash_\Sigma}$	Gen : 6.

Exercise 8.9. Show that the derivability operator $\vdash_\Sigma : \wp L_\Sigma \to \wp L_\Sigma$ has the following properties:

- extensitivity: $\Gamma \subseteq \Gamma^{\vdash_\Sigma}$;

- monotonicity: $\Gamma_1^{\vdash_\Sigma} \subseteq \Gamma_2^{\vdash_\Sigma}$ whenever $\Gamma_1 \subseteq \Gamma_2$;

- idempotence: $(\Gamma^{\vdash_\Sigma})^{\vdash_\Sigma} = \Gamma^{\vdash_\Sigma}$.

8.2   Listability of Consequence in First-Order Logic

We now state the decision problem $\text{DerHyp}_\Sigma^\Gamma$ over a signature Σ and a set $\Gamma \subseteq L_\Sigma$:

$$\text{Given } \varphi \in L_\Sigma, \text{ does } \varphi \in \Gamma^{\vdash_\Sigma}?$$

For concluding that this problem is semidecidable we need first to show that $\Gamma^{\vdash_\Sigma}$ is the least fixed point of an operator D_Σ containing $\Gamma \cup \text{Ax}_\Sigma$. In this chapter, we overload the notion of operator defined in Chapter 6, by considering as operators functions between sets of sets.

Definition 8.9. The *one-step derivation operator*

$$D_\Sigma : \wp L_\Sigma \to \wp L_\Sigma$$

is such that

$$D_\Sigma(\Psi) = \Psi \cup \{\beta : \alpha, (\alpha \supset \beta) \in \Psi\} \cup \{(\forall x \alpha) : x \in X \text{ and } \alpha \in \Psi\}$$

for every $\Psi \in \wp L_\Sigma$.

Thus, $D_\Sigma(\Psi)$ is the one step closure of Ψ by modus ponens and generalization. Observe that D_Σ should be seen as an operator from the complete lattice (partial order where all pairs of subsets have both a least upper bound and a greatest lower bound) $(\wp L_\Sigma, \subseteq)$ to itself.

Exercise 8.10. Show that operator D_Σ is extensive, monotonic but not idempotent.

The idea now is to prove that the operator D_Σ has fixed points by showing first that it preserves unions of directed families in the sense detailed below.

Definition 8.10. We say that a family $\{\Psi_e\}_{e \in E}$ such that $\Psi_e \subseteq L_\Sigma$ is *directed* whenever, for every $e', e'' \in E$, there is $e \in E$ such that $\Psi_{e'} \cup \Psi_{e''} \subseteq \Psi_e$.

In particular, if the family is closed for binary unions, then it is directed.

Proposition 8.1. If the family $\{\Psi_e\}_{e \in E}$ is directed, then

$$D_\Sigma\left(\bigcup_{e \in E} \Psi_e\right) = \bigcup_{e \in E} D_\Sigma(\Psi_e).$$

Proof. (1) $\bigcup_{e \in E} D_\Sigma(\Psi_e) \subseteq D_\Sigma(\bigcup_{e \in E} \Psi_e)$:

For each $e \in E$, $\Psi_e \subseteq \bigcup_{e \in E} \Psi_e$, and, therefore, thanks to Exercise 8.10, $D_\Sigma(\Psi_e) \subseteq D_\Sigma(\bigcup_{e \in E} \Psi_e)$. Hence, $\bigcup_{e \in E} D_\Sigma(\Psi_e) \subseteq D_\Sigma(\bigcup_{e \in E} \Psi_e)$.

(2) $D_\Sigma(\bigcup_{e \in E} \Psi_e) \subseteq \bigcup_{e \in E} D_\Sigma(\Psi_e)$:

Take $\varphi \in D_\Sigma(\bigcup_{e \in E} \Psi_e)$. There are three cases to consider:

(a) $\varphi \in \bigcup_{e \in E} \Psi_e$. Then, there exists $e \in E$ such that $\varphi \in \Psi_e$. Therefore, there is $e \in E$ such that $\varphi \in D_\Sigma(\Psi_e)$ and, so, $\varphi \in \bigcup_{e \in E} D_\Sigma(\Psi_e)$.

(b) $\varphi \in \{\beta : \alpha, \alpha \supset \beta \in \bigcup_{e \in E} \Psi_e\}$. Then, there is α such that $\alpha, \alpha \supset \varphi \in \bigcup_{e \in E} \Psi_e$. Therefore, there are $e_1, e_2 \in E$ for which $\alpha \in \Psi_{e_1}$ and $\alpha \supset \varphi \in \Psi_{e_2}$. Since $\{\Psi_e\}_{e \in E}$ is directed, there is $e \in E$ such that $\alpha, \alpha \supset \varphi \in \Psi_e$. Hence, there is $e \in E$ such that $\varphi \in D_\Sigma(\Psi_e)$, and, so, $\varphi \in \bigcup_{e \in E} D_\Sigma(\Psi_e)$.

(c) $\varphi \in \{\forall x \alpha : x \in X \text{ and } \alpha \in \bigcup_{e \in E} \Psi_e\}$. Then, there is α such that $\forall x \alpha$ is φ for some variable x and there is $e \in E$ such that $\alpha \in \Psi_e$. Thus, there is $e \in E$ such that $\forall x \alpha \in D_\Sigma(\Psi_e)$, and, so, $\varphi \in \bigcup_{e \in E} D_\Sigma(\Psi_e)$. $\square$

Note that the fact that the family is directed was only needed for dealing with modus ponens, since the set of premises is a singleton in the case of generalization. We now define what is a fixed point of D_Σ. As expected it is similar to the concept of fixed point of an operator over functions already introduced in Definition 6.8.

Definition 8.11. Let $\Psi \subseteq L_\Sigma$. A set $\Omega \subseteq L_\Sigma$ is said to be a *fixed point of D_Σ containing* Ψ whenever $D_\Sigma(\Omega) = \Omega$ and $\Psi \subseteq \Omega$. Moreover, a set Ω is the *least fixed point of D_Σ containing* Ψ whenever Ω is a fixed point of D_Σ containing Ψ and $\Omega \subseteq \Delta$ for every fixed point Δ of D_Σ containing Ψ.

Remark 8.3. Given $\Psi \subseteq L_\Sigma$ we denote by

$$\{D_\Sigma^k(\Psi)\}_{k \in \mathbb{N}}$$

the family defined inductively as follows: $D_\Sigma^0(\Psi) = \Psi$ and $D_\Sigma^{k+1}(\Psi) = D_\Sigma(D_\Sigma^k(\Psi))$.

Exercise 8.11. Given $\Psi \subseteq L_\Sigma$, show that $\{D_\Sigma^k(\Psi)\}_{k \in \mathbb{N}}$ is a directed family.

Proposition 8.2. Let $\Psi \subseteq L_\Sigma$. Then, $\bigcup_{k \in \mathbb{N}} D_\Sigma^k(\Psi)$ is the least fixed point of D_Σ containing Ψ.

Proof. The first step is to check that $\bigcup_{k \in \mathbb{N}} D_\Sigma^k(\Psi)$ is a fixed point of D_Σ containing Ψ. Clearly, $\Psi \subseteq \bigcup_{k \in \mathbb{N}} D_\Sigma^k(\Psi)$. Observe that $\{D_\Sigma^k(\Psi)\}_{k \in \mathbb{N}}$ is directed, by Exercise 8.11. Hence, by Proposition 8.1, we have:

$$D_\Sigma\left(\bigcup_{k \in \mathbb{N}} D_\Sigma^k(\Psi)\right) = \bigcup_{k \in \mathbb{N}} D_\Sigma(D_\Sigma^k(\Psi)) = \bigcup_{k \in \mathbb{N}} D_\Sigma^{k+1}(\Psi) = \bigcup_{k \in \mathbb{N}^+} D_\Sigma^k(\Psi) = \bigcup_{k \in \mathbb{N}} D_\Sigma^k(\Psi).$$

It only remains to verify that $\bigcup_{k \in \mathbb{N}} D_\Sigma^k(\Psi)$ is the least fixed point of D_Σ containing Ψ. Let Δ be a fixed point of D_Σ containing Ψ. Then,

$$\bigcup_{k \in \mathbb{N}} D_\Sigma^k(\Psi) \subseteq \Delta$$

since $D_\Sigma^k(\Psi) \subseteq \Delta$ for every $k \in \mathbb{N}$, a fact proved by induction on k as follows:

(Base) $D_\Sigma^0(\Psi) = \Psi \subseteq \Delta$.

(Step) Let $k = j + 1$. Then, $D_\Sigma^k(\Psi) = D_\Sigma(D_\Sigma^j(\Psi))$. By the induction hypothesis, $D_\Sigma^j(\Psi) \subseteq \Delta$. Therefore, by Exercise 8.10, $D_\Sigma(D_\Sigma^j(\Psi)) \subseteq D_\Sigma(\Delta)$, and, so, $D_\Sigma^k(\Psi) \subseteq \Delta$. $\qquad \square$

As a corollary of the previous proposition, the following result holds.

Proposition 8.3. Let $\Gamma \subseteq L_\Sigma$. Then, $\Gamma^{\vdash_\Sigma} = \bigcup_{k \in \mathbb{N}} D_\Sigma^k(\Gamma \cup \mathrm{Ax}_\Sigma)$.

Proof. ($\subseteq$) We prove that $\varphi \in \Gamma^{+\Sigma}$ implies $\varphi \in \bigcup_{k \in \mathbb{N}} D_{\Sigma}^{k}(\Gamma \cup \mathrm{Ax}_{\Sigma})$ for every φ, by induction.

(Base) $\varphi \in \Gamma \cup \mathrm{Ax}_{\Sigma}$. Then the thesis follows because $\Gamma \cup \mathrm{Ax}_{\Sigma} = D_{\Sigma}^{0}(\Gamma \cup \mathrm{Ax}_{\Sigma})$.

(Step) Assume that $\varphi \in \Gamma^{+\Sigma}$. We have two cases:

(1) φ is obtained from $\psi, \psi \supset \varphi \in \Gamma^{+\Sigma}$ by MP. Then, by the induction hypothesis, $\psi, \psi \supset \varphi \in \bigcup_{k \in \mathbb{N}} D_{\Sigma}^{k}(\Gamma \cup \mathrm{Ax}_{\Sigma})$. Hence, $\psi \in D_{\Sigma}^{i}(\Gamma \cup \mathrm{Ax}_{\Sigma})$ and $\psi \supset \varphi \in D_{\Sigma}^{j}(\Gamma \cup \mathrm{Ax}_{\Sigma})$ for some $i, j \in \mathbb{N}$. Then, $\psi, \psi \supset \varphi \in D_{\Sigma}^{\max\{i,j\}}(\Gamma \cup \mathrm{Ax}_{\Sigma})$ and so $\varphi \in D_{\Sigma}^{\max\{i,j\}+1}(\Gamma \cup \mathrm{Ax}_{\Sigma})$.

(2) φ is obtained from $\psi \in \Gamma^{+\Sigma}$ by Gen. Then, by the induction hypothesis, $\psi \in \bigcup_{k \in \mathbb{N}} D_{\Sigma}^{k}(\Gamma \cup \mathrm{Ax}_{\Sigma})$. Hence, $\psi \in D_{\Sigma}^{i}(\Gamma \cup \mathrm{Ax}_{\Sigma})$ for some $i \in \mathbb{N}$. Therefore, $\varphi \in D_{\Sigma}^{i+1}(\Gamma \cup \mathrm{Ax}_{\Sigma})$.

($\supseteq$) We show that $\varphi \in D_{\Sigma}^{k}(\Gamma \cup \mathrm{Ax}_{\Sigma})$ implies $\varphi \in \Gamma^{+\Sigma}$, by induction on k.

(Base) $\varphi \in D_{\Sigma}^{0}(\Gamma \cup \mathrm{Ax}_{\Sigma})$. It is enough to observe that $D_{\Sigma}^{0}(\Gamma \cup \mathrm{Ax}_{\Sigma}) = \Gamma \cup \mathrm{Ax}_{\Sigma} \subseteq \Gamma^{+\Sigma}$.

(Step) $\varphi \in D_{\Sigma}^{k+1}(\Gamma \cup \mathrm{Ax}_{\Sigma})$. There are two cases:

(1) $\psi, \psi \supset \varphi \in D_{\Sigma}^{k}(\Gamma \cup \mathrm{Ax}_{\Sigma})$. By the induction hypothesis, $\psi, \psi \supset \varphi \in \Gamma^{+\Sigma}$ and so $\varphi \in \Gamma^{+\Sigma}$ by MP.

(2) φ is $\forall x \, \psi$ and $\psi \in D_{\Sigma}^{k}(\Gamma \cup \mathrm{Ax}_{\Sigma})$. By the induction hypothesis, $\psi \in \Gamma^{+\Sigma}$ and so $\varphi \in \Gamma^{+\Sigma}$ by Gen. $\qquad\qquad\square$

We are ready to start discussing the decision problem $\mathrm{DerHyp}_{\mathrm{FOL}}$.

Exercise 8.12. Show that the set of axioms Ax_{Σ} is decidable.

In order to discuss the role of MP and Gen we introduce the following definition.

Definition 8.12. For every $\Psi \subseteq L_{\Sigma}$, let

- $\mathrm{dgen}_{\Sigma}(\Psi) = \Psi \cup \{(\forall x \, \alpha) : x \in X \text{ and } \alpha \in \Psi\}$;

- $\mathrm{dmp}_{\Sigma}(\Psi) = \Psi \cup \{\beta : \alpha, (\alpha \supset \beta) \in \Psi\}$.

Analogously to Example 1.3, we represent the first-order formula

$$\forall x \, \alpha$$

by

$$\langle \text{"forall"}, x, \beta \rangle$$

where β is the representation of α. In the sequel, we confuse $\mathrm{dgen}_{\Sigma}(\Psi)$ with the set with the representation of each formula in $\mathrm{dgen}_{\Sigma}(\Psi)$. Similarly for $\mathrm{dmp}_{\Sigma}(\Psi)$. Recall the function

$$\mathrm{univ}_{\mathrm{W}} : \mathbb{N} \times \mathrm{W} \rightharpoonup \mathrm{W}$$

introduced in Exercise 3.10, universal with respect to $\mathscr{C}_{1}^{\mathrm{W}}$.

Proposition 8.4. Let $\Psi \subseteq L_\Sigma$. There is a computable map $s_{\text{Gen}} : \mathbb{N} \to \mathbb{N}$ such that, for each $p \in \mathbb{N}$, if $(\text{univ}_W)_p$ is an enumeration of Ψ, then $(\text{univ}_W)_{s_{\text{Gen}}(p)}$ is an enumeration of $\text{dgen}_\Sigma(\Psi)$.

Proof. The result follows immediately when $\Psi = \emptyset$. Otherwise, consider the program P_{Gen^p} defined as follows:

```
function (k) (
    if k % 2 == 0 then
        a = P_K(k/2);
        b = P_L(k/2);
        return ⟨"forall", P_{h_X}(a), P_{univ_W}(p,b)⟩
    else
        return P_{univ_W}(p,(k−1)/2)
)
```

where $P_{(\text{univ}_W)_p}$, P_K, P_L and P_{h_X} are programs that compute $(\text{univ}_W)_p$, K, L and an enumeration of X, respectively. In this case, it is useful to consider strings in order to build formulas.

We now show that P_{Gen^p} computes an enumeration of $\text{dgen}_\Sigma(\Psi)$.

(1) For every $k \in \mathbb{N}$, the execution of P_{Gen^p} on k returns an element in $\text{dgen}_\Sigma(\Psi)$. Indeed:

(a) k is odd. Then the execution of $P_{(\text{univ}_W)_p}$ on $(k-1)/2$ returns $(\text{univ}_W)_p((k-1)/2) \in \Psi \subseteq \text{dgen}_\Sigma(\Psi)$.

(b) Otherwise, the execution of P_{Gen^p} on k returns $\langle\text{"forall"}, h_X(a), (\text{univ}_W)_p(b)\rangle \in \text{dgen}_\Sigma(\Psi)$ since $(\text{univ}_W)_p(b) \in \Psi$ and $h_X(a) \in X$.

(2) For every $\varphi \in \text{dgen}_\Sigma(\Psi)$, there is $k \in \mathbb{N}$ such that the execution of P_{Gen^p} on k returns φ. Assume that $\varphi \in \text{dgen}_\Sigma(\Psi)$. There are two cases to consider:

(a) $\varphi \in \Psi$. Since $(\text{univ}_W)_p$ is an enumeration of Ψ, there is m such that $(\text{univ}_W)_p(m) = \varphi$. Take $k = 2m+1$. Then, the execution of the algorithm on input k terminates with the value $(\text{univ}_W)_p(m) = \varphi$.

(b) $\varphi \in \text{dgen}_\Sigma(\Psi) \setminus \Psi$. Assume that φ is $\forall x \psi$ with $\psi \in \Psi$. Then, there are $a, b \in \mathbb{N}$ such that $h_X(a) = x$ and $(\text{univ}_W)_p(b) = \psi$. Take $k = 2J(a,b)$. Then, the execution of P_K on $k/2$ returns a and the execution of P_L on $k/2$ returns b. Moreover the execution of P_{h_X} on a returns x and the execution of $P_{(\text{univ}_W)_p}$ on b returns ψ. Thus, the execution of P_{Gen^p} on k returns φ.

Consider the program P defined as follows:

```
function (p) (return P_{J_W}(P_{Gen^p}))
```

where P_{J_W} is a program that computes J_W. It is enough to take s_{Gen} as the map computed by P. $\qquad\square$

We omit the proof of the following result because it is similar to the proof of Proposition 8.4.

Proposition 8.5. Let $\Psi \subseteq L_\Sigma$. There is a computable map $s_{MP} : \mathbb{N} \to \mathbb{N}$ such that, for each $p \in \mathbb{N}$, if $(univ_W)_p$ is an enumeration of Ψ, then $(univ_W)_{s_{MP}(p)}$ is an enumeration of $dmp_\Sigma(\Psi)$.

Observe that $D_\Sigma(\Psi) = dmp_\Sigma(\Psi) \cup dgen_\Sigma(\Psi)$.

Proposition 8.6. Let $\Psi \subseteq L_\Sigma$. There is a computable map $s_D : \mathbb{N} \to \mathbb{N}$ such that, for each $p \in \mathbb{N}$, if $(univ_W)_p$ is an enumeration of $\Psi \subseteq L_\Sigma$, then $(univ_W)_{s_D(p)}$ is an enumeration of $D_\Sigma(\Psi)$.

Proof. Consider the program P_{D^p}:

```
function (k) (
    if k%2 == 0 then
        return P_univw (P_sGen (p),k/2)
    else
        return P_univw (P_sMP (p),(k−1)/2)
)
```

where $P_{s_{Gen}}$ and $P_{s_{MP}}$ are the programs defined in the proofs of Proposition 8.4 and Proposition 8.5 for computing s_{Gen} and s_{MP}, respectively. Then P_{D^p} computes an enumeration of $D_\Sigma(\Psi)$. Consider the program P defined as follows:

$$\text{function } (p) \ (\text{return } P_{J_W}(P_{D^p}))$$

where P_{J_W} is a program that computes J_W. It is enough to take s_D as the map computed by P. $\qquad\square$

Hence, if Ψ is listable so is $D_\Sigma(\Psi)$.

Proposition 8.7. Let $\Psi \subseteq L_\Sigma$. Then $\bigcup D_\Sigma^n(\Psi)$ is listable whenever Ψ is listable.

Proof. Assume that Ψ is listable. Let $p \in \mathbb{N}$ be such that $(univ_W)_p$ is an enumeration of Ψ. Consider the following program P:

```
function (n) (
    i = 1;
    r = p;
    while i ≤ n do
        r = P_sD (r);
        i = i+1
    return r
)
```

where P_{s_D} is the program that computes s_D as defined in the proof Proposition 8.6. Observe that the execution of P on n returns an univ$_W$-index of an enumeration of $D_\Sigma^n(\Psi)$. Let s be the map computed by P. Therefore, the thesis follows from Proposition 2.10.   □

Hence, the following result holds.

Proposition 8.8. Let Σ be a FOL signature and $\Gamma \subseteq L_\Sigma$ a listable set. Then, the decision problem DerHyp$_\Sigma^\Gamma$ is semidecidable.

Proof. Let $\Gamma \subseteq L_\Sigma$ be listable. Then $\Gamma^{\vdash_\Sigma}$ is listable by Proposition 8.3 and Proposition 8.7.   □

8.3   Theories

We now give some basic notions about first-order theories.

Definition 8.13. A *theory* over a signature Σ is a set $\Theta \subseteq L_\Sigma$ such that $\Theta^{\vdash_\Sigma} = \Theta$.

As a first example, observe that, for every signature Σ, the set L_Σ is a theory (the improper one) over Σ. On the other hand, the set $\emptyset$ is not a theory.

Definition 8.14. A theory Θ over Σ is said to be *consistent* if $L_\Sigma \not\subseteq \Theta$, and is said to be *complete* if $\Theta \cap \{\varphi, (\neg \varphi)\} \neq \emptyset$ for every closed formula φ.

Note that if a theory is not consistent, then it is complete. Observe that it is possible for a theory to be consistent but not complete.

Example 8.11. For each signature Σ, theory $\emptyset^{\vdash_\Sigma}$ is consistent but not complete. On the other hand, theory L_Σ is complete and not consistent.

Definition 8.15. A theory Θ over Σ is said to be *axiomatizable* if there is a decidable set Ξ of closed formulas over Σ such that $\Xi^{\vdash_\Sigma} = \Theta$.

Unfortunately, not every theory is axiomatizable. That is the case of the theory of natural numbers as as we shall see later on.

Example 8.12. Consider the theory Θ_{RCOF} of the real closed ordered fields over the signature Σ_{RCOF} defined in Example 8.2. This theory is axiomatized by the set $\mathrm{Ax}_{\mathrm{RCOF}}$ defined as follows, where $x, y, z \in X$:

(RCOF1) $\forall x \forall y \forall z \, x + (y + z) \cong (x + y) + z$;

(RCOF2) $\forall x \, x + 0 \cong x$;

(RCOF3) $\forall x \, x + (-x) \cong 0$;

(RCOF4) $\forall x \forall y\, x + y \cong y + x$;

(RCOF5) $\forall x \forall y \forall z\, x \times (y \times z) \cong (x \times y) \times z$;

(RCOF6) $\forall x\, x \times 1 \cong x$;

(RCOF7) $\forall x \forall y \forall z\, x \times (y + z) \cong (x \times y) + (x \times z)$;

(RCOF8) $\forall x \forall y\, x \times y \cong y \times x$;

(RCOF9) $\neg\, 0 \cong 1$;

(RCOF10) $\forall x ((\neg x \cong 0) \supset \exists y\, x \times y \cong 1)$;

(RCOF11) $\forall x (\neg(x < x)$;

(RCOF12) $\forall x \forall y \forall z (x < y \supset (y < z \supset x < z))$;

(RCOF13) $\forall x \forall y (x < y \vee y < z \vee x \cong y)$;

(RCOF14) $\forall x \forall y \forall z (x < y \supset x + z < y + z)$;

(RCOF15) $\forall x \forall y ((0 < x \wedge 0 < y) \supset 0 < x \times y)$;

(RCOF16) $\forall x \exists y (y^2 \cong x \vee y^2 + x \cong 0)$;

(RCOF17) $\forall x_1 \ldots \forall x_n \, \neg (x_1^2 + \cdots + x_n^2 \cong -1)$ for each $n \in \mathbb{N}^+$;

(RCOF18) $\forall x_1 \ldots \forall x_n \exists y\, y^n + x_1 y^{n-1} + \cdots + x_{n-1} y + x_n \cong 0$ for every odd $n \in \mathbb{N}$.

The axioms from (RCOF1) to (RCOF10) are the field axioms. The axioms from (RCOF11) to (RCOF13) are the ordered field axioms. The axioms (RCOF14) to (RCOF15) state the relation between $<$ and the function symbols $+$ and $\times$, respectively. Finally, the axioms from (RCOF16) to (RCOF18) are specific for real closed fields. For more details on this theory see [40].

Since theories are sets of formulas, it makes sense to ask whether a theory is decidable or listable. First we introduce a useful notation.

Remark 8.4. Given a formula φ such that $\mathrm{fv}_\Sigma(\varphi) = \{x_1, \ldots, x_n\}$, we denote by

$$\forall \varphi$$

the closed formula $\forall x_1 \ldots \forall x_n\, \varphi$.

The following result provides a sufficient condition for a listable theory to be decidable.

Proposition 8.9. Every listable and complete theory is decidable.

Proof. Two cases are considered.

(1) If theory Θ over Σ is not consistent then $\Theta = L_\Sigma$. Hence Θ is decidable (the proof can be adapted from Exercise 2.7).

(2) Otherwise, it is still possible to give an algorithm to compute χ_Θ using the fact that Θ is listable and complete. Indeed, consider the following program:

```
function (w) (
    if P_χLΣ(w) == 0 then
        return 0;
    k = 0;
    while P_h(k) ≠ P_a(w) ∧ P_h(k) ≠ P_b(w) do
        k = k+1;
    if P_h(k) == P_a(w) then return 1 else return 0
)
```

where

- P_h is a program that computes a function h that enumerates Θ which exists because Θ is listable and non-empty;

- P_a is a program that computes $a = \lambda\, \psi\,.\,(\forall\, \psi) : L_\Sigma \to L_\Sigma$;

- P_b is a program that computes $b = \lambda\, \psi\,.\,(\neg(\forall\, \psi)) : L_\Sigma \to L_\Sigma$;

- $P_{\chi_{L_\Sigma}}$ is a program that computes $\chi_{L_\Sigma} : W \to \mathbb{N}$.

The justification of the algorithm is as follows. Observe that $w \in \Theta$ if and only if $(\forall w) \in \Theta$. Given w, assume that $w \in L_\Sigma$ (if not, the algorithm returns 0 as envisaged). Then, either $(\forall w) \in \Theta$ or $(\neg(\forall w)) \in \Theta$ but not both, since Θ is complete and consistent. Finally, because f is an enumeration of Θ there exists k such that $f(k) = (\forall w)$ if $(\forall w) \in \Theta$ or $f(k) = (\neg(\forall w))$ otherwise. In the first case the algorithm returns 1 and in the second case it returns 0. $\square$

Proposition 8.10. Every axiomatizable theory is listable.

Proof. If Θ is axiomatizable, then there exists a decidable set Ξ such that $\Xi^{\vdash_\Sigma} = \Theta$. Then, by the semidecidability of DerHyP$_\Sigma^\Xi$ (Proposition 8.8), it follows that $\Xi^{\vdash_\Sigma}$ is a listable set. $\square$

The following result is a direct corollary of Proposition 8.9 and Proposition 8.10.

Proposition 8.11. Every axiomatizable and complete theory is decidable.

Example 8.13. The theory Θ_{RCOF} is axiomatizable and complete (see [61]) and so, is decidable (Proposition 8.11).

8.4 Arithmetic

Recall the signature $\Sigma_{\mathbb{N}}$ introduced in Example 8.1.

Definition 8.16. A *theory of arithmetic* is a theory over $\Sigma_{\mathbb{N}}$ containing $\neg\, S(0) \cong 0$ as one of its theorems.

In what follows two theories of arithmetic are important. The *standard theory of arithmetic* or simply *arithmetic*, denoted by $\mathrm{Th}(\mathbb{N})$, includes all closed formulas about natural numbers that are true.

We will prove that $\mathrm{Th}(\mathbb{N})$ is not axiomatizable which is a consequence of the fact that it is possible to represent all computable maps in $\mathrm{Th}(\mathbb{N})$. Interestingly, computable maps are already representable in theories of arithmetic significantly weaker than $\mathrm{Th}(\mathbb{N})$ (a corollary of Proposition 8.19 and Proposition 8.20), namely in the theory **N**, see [58], that we now introduce.

Definition 8.17. The theory **N** is $\mathrm{Ax}_{\mathbf{N}}^{\vdash_{\Sigma_{\mathbb{N}}}}$ where $\mathrm{Ax}_{\mathbf{N}}$ is the following set of axioms:

E1 $\forall x_1\; x_1 \cong x_1$;

E2a $\forall x_1 \forall x_2\, (x_1 \cong x_2 \supset S(x_1) \cong S(x_2))$;

E2b $\forall x_1 \forall x_2 \forall x_3 \forall x_4\, (x_1 \cong x_3 \supset (x_2 \cong x_4 \supset x_1 + x_2 \cong x_3 + x_4))$;

E2c $\forall x_1 \forall x_2 \forall x_3 \forall x_4\, (x_1 \cong x_3 \supset (x_2 \cong x_4 \supset x_1 \times x_2 \cong x_3 \times x_4))$;

E3a $\forall x_1 \forall x_2 \forall x_3 \forall x_4\, (x_1 \cong x_3 \supset (x_2 \cong x_4 \supset (x_1 \cong x_2 \supset x_3 \cong x_4)))$;

E3b $\forall x_1 \forall x_2 \forall x_3 \forall x_4\, (x_1 \cong x_3 \supset (x_2 \cong x_4 \supset (x_1 < x_2 \supset x_3 < x_4)))$;

N1 $\forall x_1\; \neg\, S(x_1) \cong 0$;

N2 $\forall x_1 \forall x_2\; S(x_1) \cong S(x_2) \supset x_1 \cong x_2$;

N3 $\forall x_1\; x_1 + 0 \cong x_1$;

N4 $\forall x_1 \forall x_2\; x_1 + S(x_2) \cong S(x_1 + x_2)$;

N5 $\forall x_1\; x_1 \times 0 \cong 0$;

N6 $\forall x_1 \forall x_2\; x_1 \times S(x_2) \cong (x_1 \times x_2) + x_1$;

N7 $\forall x_1\; \neg\, x_1 < 0$;

N8 $\forall x_1 \forall x_2\; x_1 < S(x_2) \equiv (x_1 < x_2 \vee x_1 \cong x_2)$;

N9 $\forall x_1 \forall x_2\, (x_1 < x_2 \vee x_1 \cong x_2 \vee x_2 < x_1)$.

Observe that it is not the case that $\forall x\, \neg\, x \cong S(x) \in \mathbf{N}$ (a well known property of the set of natural numbers). Hence, **N** is weaker than $\mathrm{Th}(\mathbb{N})$.

8.5   Representability

In this section we concentrate on the important issue of representation of maps and sets over the natural numbers. We start by defining the class of computable maps as introduced by Kurt Gödel ([20]).

Definition 8.18. The class $\mathcal{G}$ of *Gödel maps* is the least set of maps with arguments and outputs in $\mathbb{N}$ such that:

- it contains the map $\lambda\,.\,k$, for each $k \in \mathbb{N}$;

- it contains the map $\lambda\,x\,.\,0$;

- it contains $\lambda\,x,y\,.\,x+y$, $\lambda\,x,y\,.\,x\times y$ and the characteristic map $\chi_<$ of relation $<\;\subseteq \mathbb{N}^2$;

- it contains the projections $\mathrm{proj}_{[i]}^{\mathbb{N}^n}$ for each $n,i \in \mathbb{N}^+$ with $1 \le i \le n$;

- it is closed under aggregation, composition and guarded minimization (that is, $\min_{f0}^{\le \mathbb{N}}$ over maps $f : \mathbb{N}^{n+1} \to \mathbb{N}$ such that for each $(k_1,\ldots,k_n) \in \mathbb{N}^n$ there is $k \in \mathbb{N}$ such that $f(k_1,\ldots,k_n,k) = 0$).

Moreover, for each $n \in \mathbb{N}$, a subset of $\mathbb{N}^n$ is said to be in $\mathcal{G}$ if the corresponding characteristic map belongs to $\mathcal{G}$.

We observe that the class $\mathcal{G}$ coincides with the class of all computable maps over the natural numbers (see [55]).

Remark 8.5. Given $n \in \mathbb{N}$, we denote by

$$n$$

the term $\underbrace{\mathsf{S}(\cdots \mathsf{S}(0)\cdots)}_{n}$ in $T_{\Sigma_{\mathbb{N}}}$.

Definition 8.19. Let Θ be a theory of arithmetic and $h : \mathbb{N}^n \to \mathbb{N}^m$ a map. We say that h is *represented* in Θ by $\varphi \in L_{\Sigma_{\mathbb{N}}}$ if:

- $\mathrm{fv}_{\Sigma_{\mathbb{N}}}(\varphi) = \{x_1,\ldots,x_{n+m}\}$;

- for every $a_1,\ldots,a_n,b_1,\ldots,b_m \in \mathbb{N}$,

$$\text{if } h(a_1,\ldots,a_n) = (b_1,\ldots,b_m) \text{ then } ([\varphi]_{\mathbf{a_1},\ldots,\mathbf{a_n}}^{x_1,\ldots,x_n} \equiv (\bigwedge_{i=1}^{m}(x_{n+i} \cong \mathbf{b_i}))) \in \Theta.$$

A map is said to be *representable* in Θ if it is represented in Θ by some formula.

Proposition 8.12. Every computable map in $\mathcal{G}$ is representable in **N**.

Proof. The proof is by induction on the structure of the construction certifying that the map is in $\mathcal{G}$.

(Base) There are several cases:

(1) Constant $\lambda . k$ is represented by $(x_1 \cong k)$.

(2) Map $\lambda x.0$ is represented by $((x_2 \cong \mathbf{0}) \wedge (x_1 \cong x_1))$.

(3) Each projection $\text{proj}_{[i]}^{\mathbb{N}^n}$ is represented by $((x_{n+1} \cong x_i) \wedge (x_1 \cong x_1) \wedge \ldots \wedge (x_n \cong x_n))$.

(4) Addition is represented by $(x_3 \cong (x_1 + x_2))$.

(5) Multiplication is represented by $(x_3 \cong (x_1 \times x_2))$.

(6) Map $\chi_<$ is represented by $(((x_1 < x_2) \supset (x_3 \cong \mathbf{1})) \wedge ((\neg(x_1 < x_2)) \supset (x_3 \cong \mathbf{0})))$.

(Step) There are several cases:

(1) Given $h_i : \mathbb{N}^n \to \mathbb{N}$ in $\mathcal{G}$ for $i = 1, \ldots, m$ and so, by the induction hypothesis, with each represented by some formula φ_{h_i}, the aggregation $\langle h_1, \ldots, h_m \rangle : \mathbb{N}^n \to \mathbb{N}^m$ is represented by

$$\varphi_{h_1} \wedge [\varphi_{h_2}]_{x_{n+2}}^{x_{n+1}} \wedge \ldots \wedge [\varphi_{h_m}]_{x_{n+m}}^{x_{n+1}}.$$

(2) Given $f : \mathbb{N}^n \to \mathbb{N}^m$ and $g : \mathbb{N}^m \to \mathbb{N}^r$ in $\mathcal{G}$ and so, by the induction hypothesis, represented by some formulas φ_f and φ_g, respectively, the composition $g \circ f : \mathbb{N}^n \to \mathbb{N}^r$ is represented by

$$(\exists x_{j+1} \ldots \exists x_{j+m} ([[\varphi_g]_{x_{j+1} \ldots x_{j+m}}^{x_1 \ldots x_m}]_{x_{n+1}, \ldots, x_{n+r}}^{x_{m+1}, \ldots, x_{m+r}} \wedge [\varphi_f]_{x_{j+1}, \ldots, x_{j+m}}^{x_{n+1}, \ldots, x_{n+m}}))$$

where $j = \max\{n+m, m+r, n+r\}$.

(3) Given $h : \mathbb{N}^{n+1} \to \mathbb{N}$ in $\mathcal{G}$ such that for every $(a_1, \ldots, a_n) \in \mathbb{N}^n$ there exists a for which $h(a_1, \ldots, a_n, a) = 0$, represented, by the induction hypothesis, by φ_h, the (hence guarded) minimization $\min_{h0}^{\leq \mathbb{N}} : \mathbb{N}^n \to \mathbb{N}$ is represented by

$$([\varphi_h]_0^{x_{n+2}} \wedge (\forall x_{n+3} ((x_{n+3} < x_{n+1}) \supset (\neg [\varphi_h]_{x_{n+3}0}^{x_{n+1}x_{n+2}})))).$$

The proof of this result is challenging in some cases and is outside the scope of this book. The reader should consult Section 6.7 of Shoenfields's textbook [58] for guidance. □

The concept of representability is extended to sets as follows.

Definition 8.20. Let Θ be a theory of arithmetic and $C \subseteq \mathbb{N}^n$. Then, set C is said to be *represented* in Θ by formula $\varphi \in L_{\Sigma_\mathbb{N}}$ if:

- $\text{fv}_{\Sigma_\mathbb{N}}(\varphi) = \{x_1, \ldots, x_n\}$;

- for every $a_1, \ldots, a_n \in \mathbb{N}$,

- if $(a_1, \ldots, a_n) \in C$ then $[\varphi]_{\mathbf{a_1}, \ldots, \mathbf{a_n}}^{x_1, \ldots, x_n} \in \Theta$;
- otherwise, $(\neg [\varphi]_{\mathbf{a_1}, \ldots, \mathbf{a_n}}^{x_1, \ldots, x_n}) \in \Theta$.

A set is said to be *representable* in Θ if it is represented in Θ by some formula.

Proposition 8.13. Every decidable set is representable in **N**.

Proof. We start by stating that

$$(\dagger) \quad C \text{ is representable in } \Theta \quad \text{iff} \quad \chi_C \text{ is representable in } \Theta.$$

Indeed,

($\rightarrow$) Assume that C is represented in Θ by ψ. Then χ_C is represented by the formula

$$((\psi \wedge (x_{n+1} \cong \mathbf{1})) \vee ((\neg \psi) \wedge (x_{n+1} \cong \mathbf{0}))).$$

($\leftarrow$) Assume that χ_C is represented in Θ by φ. Then, C is represented by $[\varphi]_{\mathbf{1}}^{x_{n+1}}$.
For the detailed proof of ($\dagger$) see [55].

Let D be a decidable set. Then χ_D is a computable map and so, by Proposition 8.12 is representable in **N**. Thus, by ($\dagger$), C is representable in **N**. $\square$

8.6   Gödel's First Incompleteness Theorem

In order to obtain incompleteness of arithmetic, Kurt Gödel (see [20, 21]) resorted to the *diagonalization technique* due to Georg Cantor (Cantor used this technique for the first time to show that the cardinality of the power set of any given set is strictly greater than the cardinality of the original set).

Proposition 8.14 (Cantor's Lemma). Let $C \subseteq \mathbb{N}^2$. Then, for every $D \subseteq \mathbb{N}$,

$$\lambda k . (1 - \chi_C(k, k)) \notin \{\lambda k . \chi_C(d, k) : d \in D\}.$$

That is,

$$\{k \in \mathbb{N} : k \notin C_k\} \notin \{C_d : d \in D\}$$

where C_d is the section d of C (see Definition 3.11).

The representation of arithmetical reasoning within arithmetic itself was Kurt Gödel's starting point in his way toward the incompleteness theorems. To this end, we begin by choosing a Gödelization of the relevant set of symbols.

Definition 8.21. We denote by

$$A_{\Sigma_{\mathbb{N}}}$$

the *alphabet induced by* $\Sigma_{\mathbb{N}}$, that is, the set containing the function and predicate symbols in the signature $\Sigma_{\mathbb{N}}$ (Example 8.1), the set X of variables, the connectives $\neg$

and $\supset$, the quantifier $\forall$ and the punctuation symbols. Moreover, we choose once and for all a Gödelization (see Definition 2.6)

$$g_N$$

of A_{Σ_N}.

Definition 8.22. Let $\mathrm{rsb} : N \times N \to N$ be the map defined as follows:

$$\lambda \, d, u \,. \begin{cases} g_N\left([g_N^{-1}(d)]_{g_N^{-1}(u)}^{x_1}\right) & \text{if } \begin{cases} d \in g_N(L_{\Sigma_N}) \\ u \in g_N(T_{\Sigma_N}) \end{cases} \\ d & \text{otherwise} \end{cases} .$$

Clearly, rsb brings to the realm of natural numbers the binary operation

$$\lambda \, \varphi, t \,. [\varphi]_t^{x_1} : L_{\Sigma_N} \times T_{\Sigma_N} \to L_{\Sigma_N}$$

that given a formula and a term returns the formula with the free occurrences of x_1 replaced by the term.

Definition 8.23. Given a theory of arithmetic Θ, we say that the binary relation

$$\mathrm{ths}_\Theta = \{(g_N(\delta), k) : [\delta]_{\mathbf{k}}^{x_1} \in \Theta\} \subseteq N \times N$$

is the (relational) *extent* of Θ.

Proposition 8.15. Let Θ be a theory of arithmetic. Then:

$$\mathrm{ths}_\Theta = \{(d, k) : \mathrm{rsb}(d, g_N(\mathbf{k})) \in g_N(\Theta)\}.$$

Proof. Let $S_\Theta = \{(d, k) : \mathrm{rsb}(d, g_N(\mathbf{k})) \in g_N(\Theta)\}$.

(i) $\mathrm{ths}_\Theta \subseteq S_\Theta$:

$$\begin{aligned} &\text{if} && (g_N(\delta), k) \in \mathrm{ths}_\Theta \\ &\text{then} && [\delta]_{\mathbf{k}}^{x_1} \in \Theta && (\dagger) \\ &&& g_N([\delta]_{\mathbf{k}}^{x_1}) \in g_N(\Theta) \\ &&& g_N\left([g_N^{-1}(g_N(\delta))]_{g_N^{-1}(g_N(\mathbf{k}))}^{x_1}\right) \in g_N(\Theta) \\ &&& \mathrm{rsb}(g_N(\delta), g_N(\mathbf{k})) \in g_N(\Theta) && (\ddagger) \\ &&& (g_N(\delta), k) \in S_\Theta \end{aligned}$$

(ii) $S_\Theta \subseteq \mathrm{ths}_\Theta$:

First note that if $(d, k) \in S_\Theta$ then $d \in g_N(L_{\Sigma_N})$ since otherwise $\mathrm{rsb}(d, g_N(\mathbf{k})) = d$ and, so, $\text{rsb}(d, g_{\mathbb{N}}(\mathbf{k})) \notin g_{\mathbb{N}}(\Theta)$. Thus, it is enough to prove that if $(g_{\mathbb{N}}(\delta), k) \in S_{\Theta}$ then $(g_{\mathbb{N}}(\delta), k) \in \text{ths}_{\Theta}$. Indeed:

$$
\begin{array}{ll}
\text{if} & (g_{\mathbb{N}}(\delta), k) \in S_{\Theta} \\[4pt]
\text{then} & \text{rsb}(g_{\mathbb{N}}(\delta), g_{\mathbb{N}}(\mathbf{k})) \in g_{\mathbb{N}}(\Theta) \\[4pt]
& g_{\mathbb{N}}\left([g_{\mathbb{N}}^{-1}(g_{\mathbb{N}}(\delta))]_{g_{\mathbb{N}}^{-1}(g_{\mathbb{N}}(\mathbf{k}))}^{x_1} \right) \in g_{\mathbb{N}}(\Theta) \quad (\ddagger) \\[4pt]
& g_{\mathbb{N}}([\delta]_{\mathbf{k}}^{x_1}) \in g_{\mathbb{N}}(\Theta) \\[4pt]
& [\delta]_{\mathbf{k}}^{x_1} \in \Theta \\[4pt]
& (g_{\mathbb{N}}(\delta), k) \in \text{ths}_{\Theta} \qquad\qquad\qquad\qquad\qquad (\dagger)
\end{array}
$$

where ($\dagger$) and ($\ddagger$) come as a consequence of the definitions of ths_{Θ} and rsb, respectively. $\qquad\qquad\qquad\qquad\qquad\qquad\qquad\qquad\qquad\qquad\qquad\qquad\qquad$ $\square$

The binary relation ths_{Θ} captures enough information about theory Θ for the purpose of establishing Church's Theorem by a diagonal argument. To this end, the related concept of extent of a formula in the theory of arithmetic at hand is also needed.

Definition 8.24. Given $\delta \in L_{\Sigma_{\mathbb{N}}}$, the set $\text{ext}_{\Theta}^{\delta} = \{k : [\delta]_{\mathbf{k}}^{x_1} \in \Theta\} \subseteq \mathbb{N}$. is said to be the *extent* of δ in Θ.

We omit the proof of the following result since it follows straightforwardly.

Proposition 8.16. Let Θ be a theory of arithmetic. Then:

$$
\text{ext}_{\Theta}^{\delta} = \{k : (g_{\mathbb{N}}(\delta), k) \in \text{ths}_{\Theta}\}.
$$

Proposition 8.17. Let Θ be a consistent theory of arithmetic and suppose $C \subseteq \mathbb{N}$ is represented in Θ by $\varphi \in L_{\Sigma_{\mathbb{N}}}$. Then, $C = \text{ext}_{\Theta}^{\varphi}$.

Proof. The proof is by case analysis:

$$
\begin{array}{lll}
\text{if} \quad k \in C \quad \text{then} & & \\[4pt]
& [\varphi]_{\mathbf{k}}^{x_1} \in \Theta & \text{(representability)} \\[4pt]
& k \in \text{ext}_{\Theta}^{\varphi} & \text{(definition of } \text{ext}_{\Theta}^{\varphi}); \\[8pt]
\text{if} \quad k \notin C \quad \text{then} & & \\[4pt]
& (\neg[\varphi]_{\mathbf{k}}^{x_1}) \in \Theta & \text{(representability)} \\[4pt]
& [\varphi]_{\mathbf{k}}^{x_1} \notin \Theta & \text{(consistency of } \Theta) \\[4pt]
& k \notin \text{ext}_{\Theta}^{\varphi} & \text{(definition of } \text{ext}_{\Theta}^{\varphi}).
\end{array}
$$

Therefore, $C = \text{ext}_{\Theta}^{\varphi}$. $\qquad\qquad\qquad\qquad\qquad\qquad\qquad\qquad\qquad\qquad\qquad\qquad\qquad$ $\square$

Proposition 8.18. Let Θ be a consistent theory of arithmetic where computable maps are representable and suppose $C \subseteq \mathbb{N}$ is decidable. Then,

$$C \in \{\text{ext}_\Theta^\delta : \delta \in L_{\Sigma_\mathbb{N}}\}.$$

Proof. Taking into account the hypothesis and the definition of decidable set, the decidable sets are also representable in Θ. Let φ be a formula representing C in Θ. Then, using the previous result (Proposition 8.17), $C = \text{ext}_\Theta^\varphi$. $\qquad\square$

With these results in hand, it is possible to prove the following theorem, due to Alonzo Church, that yields, in a very expedite way, Gödel's First Incompleteness Theorem, in its stronger version (toward which John Rosser also contributed).

Proposition 8.19 (Church's Theorem). Any consistent theory of arithmetic where computable maps are representable cannot be decidable.

Proof. Let Θ be a consistent theory of arithmetic where computable maps are representable. The proof uses the diagonalization technique applied to the set ths$_\Theta$. Consider the set

$$U = \{k : (k,k) \notin \text{ths}_\Theta\} \subseteq \mathbb{N}$$

with characteristic map

$$\chi_U = \lambda k . (1 - \chi_{\text{ths}_\Theta}(k,k)).$$

Observe that, thanks to Proposition 8.15,

$$(\dagger) \quad \chi_U = \lambda k . (1 - \chi_{g_\mathbb{N}(\Theta)}(\text{rsb}(k, g_\mathbb{N}(\mathbf{k})))).$$

Using Cantor's lemma, one concludes that

$$\chi_U \notin \{\lambda k . \chi_{\text{ths}_\Theta}(d,k) : d \in D\}$$

for any given $D \subseteq \mathbb{N}$. Hence, in particular,

$$\chi_U \notin \{\lambda k . \chi_{\text{ths}_\Theta}(d,k) : d \in g_\mathbb{N}(L_{\Sigma_\mathbb{N}})\}$$

and, so,

$$\chi_U \notin \{\lambda k . \chi_{\text{ths}_\Theta}(g_\mathbb{N}(\delta),k) : \delta \in L_{\Sigma_\mathbb{N}}\} \qquad \text{that is}$$
$$U \notin \{\{k : (g_\mathbb{N}(\delta),k) \in \text{ths}_\Theta\} : \delta \in L_{\Sigma_\mathbb{N}}\} \qquad \text{that is (Proposition 8.16)}$$
$$U \notin \{\text{ext}_\Theta^\delta : \delta \in L_{\Sigma_\mathbb{N}}\}$$

Hence, by Proposition 8.18, U is non-decidable and, so, χ_U is not computable. Thus, since $\lambda k . 1 - k$, rsb, $g_\mathbb{N}$ and $\lambda k . \mathbf{k}$ are computable, then, by $(\dagger)$, $\chi_{g_\mathbb{N}(\Theta)}$ cannot be computable. Thus, $g_\mathbb{N}(\Theta)$ is non-decidable and, so, by Proposition 2.16, Θ is non-decidable. $\qquad\square$

Proposition 8.20 (Gödel-Rosser's First Incompleteness Theorem). Any axiomatizable theory of arithmetic where computable maps are representable cannot be consistent and complete.

Proof. The result is reached by contradiction. Assume that there is a theory of arithmetic Θ which is complete (1), consistent (2), axiomatizable (3), and where computable maps are representable (4). Then, applying Proposition 8.10 to (3), it follows that Θ is listable (5). From (5) and (1), using Proposition 8.9, it follows that Θ is decidable (6). On the other hand, Church's Theorem, when applied to (2) and (4), yields that Θ is non-decidable, in contradiction with (6).   □

Observe that Church's Theorem and the Gödel-Rosser's First Incompleteness Theorem do not rely upon any property of $\text{Th}(\mathbb{N})$. In particular, they do not rely on the fact that computable maps are representable in $\text{Th}(\mathbb{N})$.

Profiting from the finite axiomatization of **N**, it is also shown that the set of theorems of first-order logic is not always decidable.

Proposition 8.21. No consistent extension of theory **N** is decidable.

Proof. This is a particular case of Church's Theorem (Proposition 8.19), since computable maps are representable in **N** (Proposition 8.12), and, so, in any of its extensions.   □

Proposition 8.22. The decision problem:

$$\text{Given } \varphi \in L_\Sigma, \text{ is } \varphi \text{ a theorem?}$$

is non-decidable for every signature Σ.

Proof. The proof is by contradiction. Suppose that the decision problem is decidable for every signature Σ. Then, in particular, the decision problem is decidable for $\Sigma_{\mathbb{N}}$. However, on the other hand, since the set $\text{Ax}_{\mathbb{N}}$ of specific axioms of theory **N** is finite, the following holds:

$$\varphi \in \mathbf{N} \qquad\qquad \text{iff}$$
$$\text{Ax}_{\mathbb{N}} \vdash_{\Sigma_{\mathbb{N}}} \varphi \qquad\qquad \text{iff}$$
$$\left(\bigwedge_{\delta \in \text{Ax}_{\mathbb{N}}} \delta\right) \vdash_{\Sigma_{\mathbb{N}}} \varphi \qquad\qquad \text{iff}$$
$$\vdash_{\Sigma_{\mathbb{N}}} \left(\left(\bigwedge_{\delta \in \text{Ax}_{\mathbb{N}}} \delta\right) \supset \varphi\right) \qquad\qquad \text{iff}$$
$$\left(\left(\bigwedge_{\delta \in \text{Ax}_{\mathbb{N}}} \delta\right) \supset \varphi\right) \in \emptyset^{\vdash_{\Sigma_{\mathbb{N}}}}$$

Thus, if $\emptyset^{\vdash_{\Sigma_{\mathbb{N}}}}$ were decidable, then the set **N** would also be decidable, contracting the fact that theory **N** is non-decidable (Proposition 8.21).   □

Chapter 9

Other Decision Problems

Recall Section 2.3 where decision problems are introduced. Herein, we discuss decidability of other relevant decision problems putting to good use the techniques presented in Section 2.3. We start with a very simple example.

9.1 Prime and Composite Numbers

Consider the problem of detecting prime numbers. A *prime number* is a natural number greater than or equal to 2 such that it only admits 1 and itself as divisors.

Proposition 9.1. The decision problem

Given a natural number n, is n a prime number?

is decidable.

Proof. There are several algorithms, usually called *primality tests* for this problem based on different characterization of a prime number. Let P_{fact} be the program defined in Example 1.10 that computes the factorial function. Consider the following program:

```
function (k) (
    if k ≤ 1 then
        return 0;
    if (Pfact(k − 1) + 1) % k == 0 then
        return 1
    else
        return 0
)
```

which is based on Wilson's Theorem (see [30]) that states: given $k \in \mathbb{N}$ with $k > 1$, k is a prime number iff $(k-1)! + 1 \bmod k = 0$. $\square$

Recall that a composite number is a natural number that can be obtained by multiplying two strictly smaller natural numbers greater than 1.

Proposition 9.2. The decision problem

> Given a natural number n, is n a composite number?

is decidable.

Proof. Let

$$s = \lambda n . \begin{cases} 2 & \text{if } n = 0, 1 \\ n & \text{otherwise} \end{cases}.$$

It is immediate to see that:

> n is composite   iff $s(n)$ is not prime.

Hence, by Proposition 9.1 and Proposition 2.14, the problem

> Given a natural number n, is n not a prime number?

is decidable, since s is a computable map. $\square$

9.2   Satisfiability in Propositional Logic

Recall the language L_P of propositional logic over a set P of propositional symbols introduced in Example 1.3. The semantics for propositional logic can be given by valuations.

Definition 9.1. A *valuation* is a map $v : P \to \{0, 1\}$. A valuation v *satisfies* a formula α, written $v \Vdash \alpha$ when

- $v(\alpha) = 1$, whenever $\alpha \in P$;

- $v \nVdash \beta$, whenever α is $\neg \beta$;

- either $v \nVdash \beta_1$ or $v \Vdash \beta_2$, whenever α is $\beta_1 \supset \beta_2$.

A formula α is said to be *satisfiable* if there is a valuation v such that $v \Vdash \alpha$.

Remark 9.1. We denote by $\mathrm{var}(\alpha)$ the set of propositional symbols that occur in α.

We omit the proof of the following result since it follows straightforwardly by induction.

Proposition 9.3. Let α be a formula and v_1, v_2 valuations such that $v_1(p) = v_2(p)$ for every $p \in \text{var}(\alpha)$. Then $v_1 \Vdash \alpha$ iff $v_2 \Vdash \alpha$.

Therefore, when analyzing satisfiability of a formula α it is enough to consider "restricted" valuations $v : \text{var}(\alpha) \to \{0, 1\}$.

Proposition 9.4. The decision problem

$$\text{Given a formula } \alpha \in L_P, \text{ is } \alpha \text{ satisfiable?}$$

is decidable.

Proof. We start by considering the following auxiliary program P_{set}

```
function (w) (
    i = 1;
    r = ⟨⟩;
    while i ≤ length(w) do (
        if Pcount(r, w[i]) == 0 then
            r = append(r, w[i]);
        i = i + 1
    );
    return r
)
```

that when receiving a list w returns the list with the elements of w without repetitions. Consider also the program

```
Pvar = function (a) (
    if a[1] == "p" then
        return ⟨a[2]⟩
    if a[1] == "not" then
        return Pvar(a[2])
    if a[1] == "implies" then
        return Pset(Pconc(Pvar(a[2]), Pvar(a[3])))
)
```

where P_{conc} is the program

```
function (w1, w2) (
    i = 1;
    r = w1;
    while i ≤ length(w2) do (
        r = append(r, w2[i]);
        i = i + 1
    );
    return r
)
```

Program P_{conc} returns the concatenation of two given lists w_1 and w_2 and program P_{var} given a formula a returns the list of propositional symbols in a. We now define a program P_{vals} that generates the list of all valuations for a given list of propositional symbols

```
function (n) (
    r = ⟨⟨0⟩, ⟨1⟩⟩;
    i = 2;
    while i ≤ n do (
        w = ⟨⟩;
        j = 1;
        while j ≤ length(r) do (
            w = append(v, append(r[j], 0));
            w = append(v, append(r[j], 1));
            j = j + 1
        );
        r = w;
        i = i + 1
    );
    return r
)
```

Finally consider the following auxiliary program that given a valuation and a formula returns 1 when the valuation satisfies the formula and returns 0 otherwise:

```
P_eval = function (v, u, a) (
    if a[1] == "p" then
        return v[pos(u, a[2])]
    if a[1] == "not" then
        return 1 − P_eval(v, u, a[2])
    if a[1] == "implies" then
        if P_eval(v, u, a[2]) == 1 ∧ P_eval(v, u, a[3]) == 0 then
            return 0
        else
            return 1
)
```

We are ready to define program P_{sat}

```
function (a) (
    u = P_var(a);
    w = P_vals(length(u));
    i = 1;
    r = false;
    while i ≤ length(w) ∧ ¬r do
        if P_eval(w[i],u,a) == 1 then
            r = true
        else
            i = i+1;
    return r
)
```

that when receiving a formula returns true if there is a valuation that satisfies the formula and false otherwise. Hence the satisfiability problem is decidable. □

9.3 Consequence in Propositional Logic

Recall once again the language L_P of propositional logic over a set P of propositional symbols introduced in Example 1.3 and the concept of valuation presented in Definition 9.1.

Definition 9.2. Let $\Gamma \subseteq L_P$ and $\alpha \in L_P$. We say that α is a *semantic consequence* of Γ, written $\Gamma \vDash \alpha$, whenever, for every valuation v, $v \Vdash \alpha$ when $v \Vdash \gamma$ for every $\gamma \in \Gamma$.

Remark 9.2. Given $\Gamma \subseteq L_P$, we denote by $\Gamma^{\vDash}$ the set $\{\beta \in L_P : \Gamma \vDash \beta\}$.

Proposition 9.5. The decision problem

$$\text{Given } \alpha \in L_P, \text{ does } \alpha \in \Gamma^{\vDash}?$$

is non-decidable for every set of propositional symbols P and $\Gamma \subseteq L_P$, even when Γ is a decidable set.

Proof. Let P be the set

$$\{\pi_{px} : p,x \in \mathbb{N}\} \cup \{\hat{\pi}_{px}^t : p,x,t \in \mathbb{N}\},$$

of propositional symbols and Γ the set

$\{\hat{\pi}_{px}^t : p,x,t \in \mathbb{N} \text{ with } p_W \text{ halts on input } x \text{ within } t \text{ time units with a natural number}\}$

$$\cup$$

$$\{\hat{\pi}_{px}^t \supset \pi_{px} : p,x,t \in \mathbb{N}\}.$$

It is straightforward to prove that Γ is decidable. We show that the Halting decision problem (Definition 3.5) can be reduced (Definition 2.5) to problem above.

(1) It is easy to see that the map

$$s : (p,x) \mapsto \pi_{px}$$

is computable.

(2) Reduction. We must prove that

$$(p,x) \in \text{Halting} \quad \text{iff} \quad \pi_{px} \in \Gamma^{\vDash}.$$

That is,

$$p_{\mathsf{W}} \text{ halts on input } x \text{ with a natural number} \quad \text{iff} \quad \pi_{px} \in \Gamma^{\vDash}.$$

($\rightarrow$) Assume that program p_{W} halts on input x with a natural number. Then, there is $t \in \mathbb{N}$ such that program p_{W} halts on input x within t time units with a natural number. That is, there is $t \in \mathbb{N}$ such that $\hat{\pi}_{px}^{t} \in \Gamma$. Hence, $\pi_{px} \in \Gamma^{\vDash}$ since $(\hat{\pi}_{px}^{t} \supset \pi_{px}) \in \Gamma$.

($\leftarrow$) The proof is by contraposition. If program p_{W} does not halt on input x with a natural number then, for every $t \in \mathbb{N}$, program p_{W} does not halt on input x within t time units with a natural number. That is, for every $t \in \mathbb{N}$, $\hat{\pi}_{px}^{t} \notin \Gamma$. We show that there is $v : P \to \{0,1\}$ such that

$$\begin{cases} v \Vdash \Gamma \\ v \nVdash \pi_{px} \end{cases}.$$

Just take v such that

$$v(\hat{\pi}_{qy}^{u}) = \begin{cases} 1 & \text{if } \hat{\pi}_{qy}^{u} \in \Gamma \\ 0 & \text{otherwise} \end{cases}$$

and

$$v(\pi_{qy}) = \begin{cases} 1 & \text{if there is } u \text{ such that } \hat{\pi}_{qy}^{u} \in \Gamma \\ 0 & \text{otherwise} \end{cases}.$$

Therefore,

$$\pi_{px} \notin \Gamma^{\vDash}.$$

Thus, the given decision problem is non-decidable by Proposition 2.14 since s is a computable map and because Halting is non-decidable by Proposition 3.9.　　□

9.4　Colouring and Bipartiteness of Finite Graphs

We start by introducing the notions of graph and colouring (see [6]).

Definition 9.3. A *finite graph* is a pair $G = (U,E)$ where U is a finite set (each element is a vertex) and $E \subseteq U \times U$ (each element of E is an edge) is a symmetric relation.

Definition 9.4. Let $G = (U, E)$ be a finite graph, $k \in \mathbb{N}^+$ and $\{1, \ldots, k\}$ a finite set. A *k-colouring* of G is a map

$$\xi : U \to \{1, \ldots, k\}$$

such that if $(u_1, u_2) \in E$, with $u_1 \neq u_2$, then $\xi(u_1) \neq \xi(u_2)$. We say that a graph G is *k-colourable* if there is a k-colouring of G.

Recall once again the language L_P of propositional logic over a set P of propositional symbols introduced in Example 1.3 and the concept of valuation presented in Definition 9.1.

Remark 9.3. Given a finite set $\Gamma = \{\gamma_1, \ldots, \gamma_n\}$ of propositional formulas, we denote by

$$\bigwedge \Gamma \text{ the propositional formula } \gamma_1 \wedge \ldots \wedge \gamma_n.$$

Proposition 9.6. The decision problem

Given a finite graph G and $k \in \mathbb{N}^+$, is G k-colourable?

is decidable.

Proof. We show that the decision problem at hand can be reduced to the satisfiability decision problem presented in Section 9.2. To this end, we associate with each finite graph $G = (U, E)$ and $k \in \mathbb{N}^+$, the set of propositional symbols

$$P_{G,k} = \{\pi_{uj} : u \in U, j \in \{1, \ldots, k\}\}$$

(1) Consider the map $s : (G, k) \mapsto \bigwedge \Gamma_{Gk}$ where Γ_{Gk} is the following set of formulas:

- $\displaystyle \bigwedge_{u \in U} \bigvee_{j=1}^{k} \pi_{uj}$;

- $\displaystyle \bigwedge_{u \in U} \bigwedge_{j_1, j_2 \in \{1, \ldots, k\}, j_1 \neq j_2} \neg(\pi_{uj_1} \wedge \pi_{uj_2})$;

- $\displaystyle \bigwedge_{j \in \{1, \ldots, k\}} \bigwedge_{(u_1, u_2) \in E} \neg(\pi_{u_1 j} \wedge \pi_{u_2 j})$.

The first formula in Γ_{Gk} asserts that each vertex should have a colour. The second formula states every vertex cannot be coloured with more than one colour. The last formula expresses the fact that no pair of adjacent vertexes can have the same colour. Observe that s is computable.

(2) Reduction. We must prove that

a finite graph G is k-colourable   iff $\bigwedge \Gamma_{Gk}$ is satisfiable.

($\rightarrow$) Assume that G is k-colourable. Let ξ be a k-colouring. Define the map $v_\xi :$ $P_{G,k} \rightarrow \{0,1\}$ as follows:

$$v_\xi(\pi_{uj}) = \begin{cases} 1 & \text{whenever } \xi(u) = j \\ 0 & \text{otherwise.} \end{cases}$$

(a) v_ξ is a valuation. Indeed, v_ξ is a map since ξ is a map.

(b) $v \Vdash \gamma$ for every $\gamma \in \Gamma_{Gk}$ since ξ is a k-colouring.

($\leftarrow$) Assume that Γ_{Gk} is satisfiable. Let v a valuation that satisfies Γ_{Gk}. Define ξ_v as follows:

$$\xi_v(u) = i \text{ whenever } v \Vdash \pi_{ui}.$$

We now show that ξ_v is a k-colouring of G.

(a) ξ_v is a map. Indeed, $\xi_v(u)$ is always defined since $v \Vdash \bigvee_{j=1}^k \pi_{uj}$. Moreover, $\xi_v(u)$ has just one value since $v \Vdash \neg(\pi_{uj_1} \wedge \pi_{uj_2})$.

(b) Assume that $(u_1, u_2) \in E$. Suppose, by contradiction, that

$$\xi_v(u_1) = \xi_v(u_2) = i.$$

So, by definition of ξ_v, $v \Vdash \pi_{u_1 i}$ and $v \Vdash \pi_{u_2 i}$. Furthermore, since v satisfies all the formulas in Γ_{Gk}, then $v \Vdash \neg(\pi_{u_1 i} \wedge \pi_{u_2 i})$ for every $(u_1, u_2) \in E$. That is, either $v \nVdash \pi_{u_1 i}$ or $v \nVdash \pi_{u_2 i}$ thus getting a contradiction.

Thus, the given decision problem is decidable by Proposition 2.14 since s is a computable map and the satisfiability problem is decidable by Proposition 9.4.   □

Now we want to address a decision problem associated with bipartite graphs ([6]).

Definition 9.5. A finite graph is said to be *irreflexive* if there is no edge from v to v, for every $v \in V$. An irreflexive finite graph $G = (V, E)$ is said to be *bipartite* if there is a partition $\{V_1, V_2\}$ of V (that is, $V_1 \cap V_2 = \emptyset$ and $V_1 \cup V_2 = V$) such that for every edge in E one of the endpoints is in V_1 and the other endpoint is in V_2.

Proposition 9.7. The decision problem

Given a finite graph G, is G a bipartite graph?

is decidable.

Proof. It is well known that a graph G is bipartite iff G admits a 2-colouring (see [3]). Hence, by Proposition 9.6, we can conclude that the given decision problem is decidable.   □

9.5 Euclidean Geometry

We start by defining Euclidean geometry as a FOL theory Θ_{EG} following the work of Alfred Tarski, see [62], also taking into account the work in [24]. This means that we axiomatize the relationships between points, lines, segments and rays by choosing appropriate predicate symbols. The goal is to prove that Θ_{EG} is decidable. Alfred Tarski was the first to provide a decidable FOL axiomatization of Euclidean geometry, see [62] (a second-order axiomatization of Euclidean geometry had been proposed by David Hilbert, see [27]). His proof relies on the reduction to the theory of real closed ordered fields which was shown to be decidable in [61].

Definition 9.6. Let $\Sigma_{EG} = (F, P, \tau)$ be the signature of Euclidean geometry where $F = \emptyset$, $P = \{\cong, B, E\}$, $\tau(\cong) = 2$, $\tau(B) = 3$ and $\tau(E) = 4$.

Predicate symbols B and E are interpreted as "between" and "equidistance", respectively. Hence, the atomic formula $B(x, y, z)$ means that y is between x and z, that is y is in the segment xz. Moreover, the atomic formula $E(x, y, z, w)$ is interpreted as the distance from x to y is the same as the distance from z to w, that is, the length of the segment xy is equal to the length of the segment zw.

Definition 9.7. The theory Θ_{EG} is composed by the following axioms where we omit the universal closure in all formulas:

(EG1) $E(x, y, y, x)$;

(EG2) $(E(x, y, u, v) \wedge E(u, v, r, w)) \supset E(x, y, r, w)$;

(EG3) $E(x, y, z, z) \supset (x \cong y)$;

(EG4) $\exists w(B(x, y, w) \wedge E(y, w, u, v))$;

(EG5) $((\neg x \cong y) \wedge B(x, y, z) \wedge B(x', y', z') \wedge E(x, y, x', y') \wedge E(y, z, y', z')$
$\wedge E(x, u, x', u') \wedge E(y, u, y', u')) \supset E(z, u, z', u')$;

(EG6) $B(x, y, x) \supset x \cong y$;

(EG7) $(B(x, u, z) \wedge B(y, v, z)) \supset \exists w(B(u, w, y) \wedge B(v, w, x))$;

(EG8) $\exists x \exists y \exists z((\neg B(x, y, z)) \wedge (\neg B(y, z, x)) \wedge \neg B(z, x, y))$;

(EG9) $((\neg u \cong v) \wedge E(x, u, x, v) \wedge E(y, u, y, v) \wedge E(z, u, z, v)) \supset$
$(B(x, y, z) \vee B(y, z, x) \vee B(z, x, y))$

(EG10) $(B(x, u, v) \wedge B(y, u, z) \wedge (\neg x \cong u)) \supset$
$\exists w \exists w'(B(x, y, w) \wedge B(x, z, w') \wedge B(w', v, w))$;

(EG11) $(\exists w \forall x \forall y((\varphi \wedge \psi) \supset B(w, x, y))) \supset \exists w' \forall x \forall y((\varphi \wedge \psi) \supset B(x, w', y))$ such that w, w', y do not occur free in φ and w, w', x do not occur free in ψ.

Axioms (EG1), (EG2) and (EG3) are called *reflexivity*, *transitivity* and *identity* of E and Axiom (EG6) the *identity* of B. Axiom (EG4), known as *segment construction* axiom, states that if uv is a segment and x is the starting point of a ray and y belongs to that ray then there is w such that segment uv is equidistant to segment yw. Axiom (EG5), known as *five segment* axiom, means the following: if we have two pairs of five segments (xy, yz, xu, yu, zu) and $(x'y', y'z', x'u', y'u', z'u')$ if the kth segments of the pairs have the same length for $k = 1, \ldots, 4$ then the 5th segments have also the same length. Axiom (EG7), called *Pasch* axiom, means that the two diagonals, from x to v and from u to y, of the polygon based on points x, u, v, y must intersect at some point w. Axiom (EG8) is known as *lower 2-dimensional* axiom. The meaning of this axiom is that there are three noncolinear points. Axiom (EG9), called *upper 2-dimensional* axiom, imposes that if three points are equidistant from two different points then they must be colinear. Axiom (EG10), known as *Euclid's* axiom, states that if v is in the angle $\angle yxz$ then there exists points w, w' such that the line determined by w, w' intersects both sides of the angle. Finally, we explain the meaning of Axiom (EG11) called *continuity axiom*. Assume that φ and ψ are formulas in Σ_{EG} satisfying the conditions of (EG11) such that: (1) φ and ψ define two sets of points, say P_φ and P_ψ, respectively; and (2) there is a ray with left hand point w such that every x in P_φ is always on the left of every y in P_ψ. Then, there is w' that separates x from y for every x and y in P_φ and P_ψ, respectively (it is interesting to note the relationship of this axiom to the Dedekind cuts for defining the real closed field $\mathbb{R}$ from the field $\mathbb{Q}$, see [54]).

Example 9.1. Recall the signature Σ_{EG} in Definition 9.6. Then

$$I^{EG} = (\mathbb{R}^2, \emptyset, \{\cong^{I^{EG}}, B^{I^{EG}}, E^{I^{EG}}\})$$

where

- $\cong^{I^{EG}}: \mathbb{R}^2 \times \mathbb{R}^2 \to \{0, 1\}$ is such that $\cong^{I^{EG}}((c_1, c_2), (d_1, d_2)) = 1$ iff c_1 is d_1 and c_2 is d_2;

- $B^{I^{EG}}: \mathbb{R}^2 \times \mathbb{R}^2 \times \mathbb{R}^2 \to \{0, 1\}$ is such that $B^{I^{EG}}(c, d, e) = 1$ iff c, d, e are collinear (see [16]), that is, using the slope formula,

$$(d_2 - c_2)(e_1 - d_1) = (e_2 - d_2)(d_1 - c_1)$$

and d is in between c and e, that is,

$$(c_1 - d_1)(d_1 - e_1) \geq 0 \text{ and } (c_2 - d_2)(d_2 - e_2) \geq 0;$$

- $E^{I^{EG}}: \mathbb{R}^2 \times \mathbb{R}^2 \times \mathbb{R}^2 \times \mathbb{R}^2 \to \{0, 1\}$ is such that $E^{I^{EG}}(c, d, c', d') = 1$ iff

$$(d_1 - c_1)^2 + (d_2 - c_2)^2 = (d'_1 - c'_1)^2 + (d'_2 - c'_2)^2;$$

is an interpretation structure for Σ_{EG}.

Example 9.2. Recall the signature Σ_{EG} in Definition 9.6 and the interpretation structure I^{EG} in Example 9.1. Then, for instance $I^{EG}\rho \Vdash E(x,y,z,z) \supset x \cong y$ for every assignment ρ.

It is not difficult to see that every interpretation structure for Θ_{EG} is based on an interpretation structure for Θ_{RCOF} as we describe now.

Definition 9.8. Let $I = (D, \{0^I, 1^I, -^I, +^I, \times^I\}, \{\cong^I, <^I\})$ be an interpretation structure for Θ_{RCOF}. The *Euclidean geometry interpretation structure*

$$EG(I) = (D^2, \emptyset, \{\cong^{EG(I)}, B^{EG(I)}, E^{EG(I)}\})$$

induced by I is such that:

- $\cong^{EG(I)}: D^2 \times D^2 \to \{0,1\}$ is the map such that: $\cong^{EG(I)} (c,d) = 1$ iff $c_1 \cong^I d_1$ and $c_2 \cong^I d_2$;

- $B^{EG(I)} : D^2 \times D^2 \times D^2 \to \{0,1\}$ is the map such that:

$$B^{EG(I)}(c,d,e) = 1$$

iff

$$\begin{cases} (d_2 +^I (-^I c_2)) \times^I (e_1 +^I (-^I d_1)) = (e_2 +^I (-^I d_2)) \times^I (d_1 +^I (-^I c_1)) \\ 0^I \leq^I (c_1 +^I (-^I d_1)) \times^I (d_1 +^I (-^I e_1)) \\ 0^I \leq^I (c_2 +^I (-^I d_2)) \times^I (d_2 +^I (-^I e_2)) \end{cases} ;$$

- $E^{EG(I)} : D^2 \times D^2 \times D^2 \times D^2 \to \{0,1\}$ is the map such that:

$$E^{EG(I)}(c,d,c',d') = 1$$

iff

$$(d_1 +^I (-^I c_1))^2 +^I (d_2 +^I (-^I c_2))^2 = (d'_1 +^I (-^I c'_1))^2 +^I (d'_2 +^I (-^I c'_2))^2.$$

The decidability of the FOL theory Θ_{EG} of Euclidean geometry is obtained by reduction to Θ_{RCOF} following [61] since Θ_{RCOF} is decidable (Example 8.13).

Definition 9.9. Let

$$s_{EG \to RCOF} : L_{\Sigma_{EG}} \to L_{\Sigma_{RCOF}}$$

be the *translation map* defined inductively as follows:

- $s_{EG \to RCOF}(x \cong y)$ is the formula $(x_1 \cong y_1) \wedge (x_2 \cong y_2)$;

- $s_{EG \to RCOF}(B(x,y,z))$ is the formula

$$(y_2 + (-x_2)) \times (z_1 + (-y_1)) \cong (z_2 + (-y_2)) \times (y_1 + (-x_1)) \wedge$$
$$(0 < (x_1 + (-y_1)) \times (y_1 + (-z_1)) \vee (0 \cong (x_1 + (-y_1)) \times (y_1 + (-z_1))) \wedge$$
$$(0 < (x_2 + (-y_2)) \times (y_2 + (-z_2)) \vee (0 \cong (x_2 + (-y_2)) \times (y_2 + (-z_2)));$$

- $s_{EG \to RCOF}(E(x,y,z,w))$ is the formula

$$(x_1 + (-y_1))^2 + (x_2 + (-y_2))^2 \cong (z_1 + (-w_1))^2 + (z_2 + (-w_2))^2;$$

- $s_{EG \to RCOF}(\neg \varphi)$ is the formula $\neg s_{EG \to RCOF}(\varphi)$;

- $s_{EG \to RCOF}(\varphi_1 \supset \varphi_2)$ is the formula $s_{EG \to RCOF}(\varphi_1) \supset s_{EG \to RCOF}(\varphi_2)$;

- $s_{EG \to RCOF}(\forall x \varphi)$ is the formula $\forall x_1 \forall x_2 \, s_{EG \to RCOF}(\varphi)$.

The translation of $B(x,y,z)$ reflects that if (x_1, x_2), (y_1, y_2) and (z_1, z_2) are collinear then the slope of the lines determined by $(x_1, x_2), (y_1, y_2)$ and by $(y_1, y_2), (z_1, z_2)$ should be the same. Moreover, the translation of $B(x,y,z)$ expresses that y is in between x and z. The translation of $E(x,y,z,w)$ states that the distance from x to y should be the same as the distance from z to w. We are ready to prove that Θ_{EG} is decidable.

Proposition 9.8. Theory Θ_{EG} is decidable.

Proof. The first step is to prove that for each interpretation structure I for Θ_{RCOF}, the following statement holds:

$$EG(I) \Vdash \varphi \quad \text{iff} \quad I \Vdash s_{EG \to RCOF}(\varphi)$$

for every $\varphi \in L_{\Sigma_{EG}}$ by induction on the structure of φ (we omit the proof since it follows straightforwardly). Observe that the base step follows almost immediately taking into account the translation map on atomic formulas. Hence

$$\Theta_{EG} \vDash \varphi \quad \text{iff} \quad \Theta_{RCOF} \vDash s_{EG \to RCOF}(\varphi)$$

since all the models of Θ_{EG} are induced by the models of Θ_{RCOF}. Thus, the map $s_{EG \to RCOF}$ reduces the problem

Given $\varphi \in L_{\Sigma_{EG}}$, does $\varphi \in \Theta_{EG}$?

to the problem

Given $\psi \in L_{\Sigma_{RCOF}}$, does $\psi \in \Theta_{RCOF}$?

The thesis follows by Proposition 2.14 since $s_{EG \to RCOF}$ is a computable map and Θ_{RCOF} is decidable (see Example 8.13). $\qquad\square$

9.6 Modal Logic

We show that modal logic K is decidable by reducing K to a decidable fragment of FOL (see [22]). We start with a very brief introduction to modal logic (see [5]).

Definition 9.10. Given a set P of propositional symbols, the set L_{ML}^P of *modal formulas* over P is inductively defined as follows:

- $P \subseteq L_{ML}^P$;

- $\neg \varphi, \Diamond \varphi \in L_{ML}^P$, provided that $\varphi \in L_{ML}^P$;

- $\varphi_1 \supset \varphi_2 \in L_{ML}^P$, provided that $\varphi_1, \varphi_2 \in L_{ML}^P$.

The elements of P are *propositional symbols*. A formula $\Diamond \varphi$ means *possibly* φ. Besides the usual abbreviations $\varphi_1 \wedge \varphi_2$, $\varphi_1 \vee \varphi_2$ and $\varphi_1 \equiv \varphi_2$ we also have $\Box \varphi$ which is $\neg \Diamond \neg \varphi$ meaning *necessarily* φ. The semantics is herein introduced by Kripke structures.

Definition 9.11. A *Kripke structure* over P for modal logic is a triple

$$(W, R, V)$$

where

- W is a non-empty set whose elements are called *worlds*;

- $R \subseteq W^2$ is a relation called the *accessibility relation*;

- $V : P \to \wp W$ is a map called *valuation*.

We denote by KS^P the class of all Kripke structures over P for modal logic.

Definition 9.12. The *local satisfaction* of a formula φ by a Kripke structure (W, R, V) at $w \in W$, written

$$(W, R, V), w \Vdash \varphi$$

is inductively defined as follows:

- $(W, R, V), w \Vdash p$ if $w \in V(p)$ for $p \in P$;

- $(W, R, V), w \Vdash \neg \varphi$ if $(W, R, V), w \nVdash \varphi$;

- $(W, R, V), w \Vdash \varphi_1 \supset \varphi_2$ if either $(W, R, V), w \nVdash \varphi_1$ or $(W, R, V), w \Vdash \varphi_2$;

- $(W, R, V), w \Vdash \Diamond \varphi$ if $(W, R, V), w' \Vdash \varphi$ for some $w' \in W$ such that wRw'.

A formula φ is *satisfiable* if there are a Kripke structure (W, R, V) and $w \in W$ such that $(W, R, V), w \Vdash \varphi$.

Example 9.3. The formula $(\Diamond(\varphi_1 \vee \varphi_2)) \equiv ((\Diamond\varphi_1) \vee (\Diamond\varphi_2))$ is locally satisfied by every Kripke structure (W,R,V) and $w \in W$.

We now define the so called standard translation (see [5]) from modal logic to first-order logic (the basic concepts of first-order logic can be seen in Section 8.1).

Definition 9.13. Let Σ_{ML}^P be the FOL signature induced by modal logic over P with no function symbols, set of predicate symbols $\{\bar{p} : p \in P\}$ with arity 1 and predicate symbol $\bar{R}$ of arity two.

Remark 9.4. We denote by $\mathrm{IS}_{\Sigma_{\mathrm{ML}}^P}$ the class of all FOL interpretation structures over Σ_{ML}^P.

Definition 9.14. Given $x,y \in X$, define the *standard translation maps*

$$\mathrm{ST}_x, \mathrm{ST}_y : L_{\mathrm{ML}}^P \to L_{\Sigma_{\mathrm{ML}}^P}$$

inductively as follows:

- $\mathrm{ST}_x(p) = \bar{p}(x)$ and $\mathrm{ST}_y(p) = \bar{p}(y)$;
- $\mathrm{ST}_x(\neg\varphi) = \neg\mathrm{ST}_x(\varphi)$ and $\mathrm{ST}_y(\neg\varphi) = \neg\mathrm{ST}_y(\varphi)$;
- $\mathrm{ST}_x(\varphi_1 \supset \varphi_2) = \mathrm{ST}_x(\varphi_1) \supset \mathrm{ST}_x(\varphi_2)$ and $\mathrm{ST}_y(\varphi_1 \supset \varphi_2) = \mathrm{ST}_y(\varphi_1) \supset \mathrm{ST}_x(\varphi_2)$;
- $\mathrm{ST}_x(\Diamond\varphi) = \exists y(\bar{R}(x,y) \wedge \mathrm{ST}_y(\varphi))$ and $\mathrm{ST}_y(\Diamond\varphi) = \exists x(\bar{R}(y,x) \wedge \mathrm{ST}_x(\varphi))$.

The standard translation induces a relationship between Kripke structures over P for modal logic and FOL interpretation structures over Σ_{ML}^P.

Definition 9.15. The *semantical translation map*

$$\mathrm{ST} : \mathrm{KS}^P \to \mathrm{IS}_{\Sigma_{\mathrm{ML}}^P}$$

is defined as follows:

$$\mathrm{ST}(W,R,V) = (W, \{\bar{p}^P\}_{p \in P}, \bar{R}^P)$$

where

- $\bar{p}^P(w) = 1$ iff $w \in V(p)$;
- $\bar{R}^P(w_1,w_2) = 1$ iff $w_1 R w_2$.

We now show that local satisfaction carries over from modal logic to first-order logic.

Proposition 9.9. Let $(W,R,V) \in KS^P$, $w, w' \in W$ and φ a modal formula over P. Then,

$$(W,R,V), w \Vdash \varphi \text{ iff } ST(W,R,V), \rho \Vdash ST_x(\varphi)$$

where ρ is an assignment such that $\rho(x) = w$ and

$$(W,R,V), w' \Vdash \varphi \text{ iff } ST(W,R,V), \rho' \Vdash ST_y(\varphi)$$

where ρ' is an assignment such that $\rho'(y) = w'$.

Proof. We prove the result by induction on φ.

(Base) Let φ be $p \in P$. Hence, $(W,R,V), w \Vdash p$ iff $w \in V(p)$ iff $\bar{p}^P(w) = 1$ iff $ST(W,R,V), \rho \Vdash_\Sigma \bar{p}(x)$ iff $ST(W,R,V), \rho \Vdash ST_x(p)$.

(Step) We have three cases:

(a) φ is $\neg \psi$. Then, $(W,R,V), w \Vdash \neg \psi$ iff $(W,R,V), w \not\Vdash \psi$ iff (by the induction hypothesis, $ST(W,R,V), \rho \not\Vdash ST_x(\psi)$ iff $ST(W,R,V), \rho \Vdash ST_x(\neg \psi)$.

(b) φ is $\psi_1 \supset \psi_2$. Similar to (a).

(c) φ is $\Diamond \psi$. We start by showing that

$$(W,R,V), w \Vdash \Diamond \psi \quad \text{implies} \quad ST(W,R,V), \rho \Vdash ST_x(\Diamond \psi).$$

Assume that $(W,R,V), w \Vdash \Diamond \psi$. Then, there is $w' \in W$ with wRw' and $(W,R,V), w' \Vdash \psi$. We must show that

$$ST(W,R,V), \rho \Vdash \exists y(\bar{R}(x,y) \wedge ST_y(\psi)).$$

Let $\rho' \equiv_y \rho$ be such that $\rho'(y) = w'$. Then,

$$ST(W,R,V), \rho' \Vdash \bar{R}(x,y)$$

since $\bar{R}^P(w,w') = 1$. Moreover,

$$ST(W,R,V), \rho' \Vdash ST_y(\psi)$$

by the induction hypothesis, since $(W,R,V), w' \Vdash \psi$. We now prove that

$$ST(W,R,V), \rho \Vdash ST_x(\Diamond \psi) \quad \text{implies} \quad (W,R,V), w \Vdash \Diamond \psi.$$

Assume that $ST(W,R,V), \rho \Vdash ST_x(\Diamond \psi)$. Hence

$$ST(W,R,V), \rho \Vdash \exists y(\bar{R}(x,y) \wedge ST_y(\psi)).$$

Thus, there is $\rho' \equiv_y \rho$ such that

$$(\dagger) \quad ST(W,R,V), \rho' \Vdash \bar{R}(x,y)$$

and

$$(\ddagger) \quad \mathrm{ST}(W,R,V),\rho' \Vdash \mathrm{ST}_y(\psi).$$

From $(\dagger)$, we conclude that $\rho'(x)R\rho'(y)$ in (W,R,V) and from $(\ddagger)$ we can conclude, by the induction hypothesis, that $(W,R,V),\rho'(y) \Vdash \psi$. Therefore, $(W,R,V),w \Vdash \Diamond\psi$ because $\rho'(x) = \rho(x) = w$. $\square$

On the other hand, we have to define another map from FOL interpretation structures over Σ_{ML}^P to the Kripke structures over P.

Definition 9.16. The *reverse semantical translation map*

$$\mathrm{RT} : \mathrm{IS}_{\Sigma_{\mathrm{ML}}^P} \to \mathrm{KS}^P$$

is defined as follows

$$\mathrm{RT}(W, \{\bar{p}^P\}_{p \in P}, \bar{R}^P) = (W,R,V)$$

where

- $R = \{(w_1, w_2) \in W^2 : \bar{R}^P(w_1, w_2) = 1\}$;

- $V(p) = \{w \in W : \bar{p}^P(w) = 1\}$.

The following result shows that RT preserves and reflects local satisfaction. We omit its proof since it follows the same steps as the proof of Proposition 9.9.

Proposition 9.10. Let $I \in \mathrm{IS}_{\Sigma_{\mathrm{ML}}^P}$, ρ and ρ' assignments over I and φ a modal formula over P. Then,

$$I,\rho \Vdash \mathrm{ST}_x(\varphi) \quad \text{iff} \quad \mathrm{RT}(I),\rho(x) \Vdash \varphi$$

and

$$I,\rho' \Vdash \mathrm{ST}_y(\varphi) \quad \text{iff} \quad \mathrm{RT}(I),\rho'(y) \Vdash \varphi.$$

Proposition 9.11. The satisfiability problem in modal logic K is decidable.

Proof. (1) Let φ be a modal formula. Assume that $\mathrm{ST}_x(\varphi)$ is not satisfiable in FOL. Then, there are no FOL interpretation structure I over Σ_{ML}^P and assignment ρ such that

$$I,\rho \Vdash \mathrm{ST}_x(\varphi).$$

In particular, there are no Kripke structure (W,R,V) such that

$$\mathrm{ST}(W,R,V),\rho \Vdash \mathrm{ST}_x(\varphi).$$

Hence, by Proposition 9.9, there is no Kripke structure (W,R,V) and $w \in W$ such that $(W,R,V),w \Vdash \varphi$ and so φ is not satisfiable in K modal logic.

(2) Let φ be a modal formula. Assume that $ST_x(\varphi)$ is satisfiable in FOL. Then, there are interpretation structure I over Σ and assignment ρ such that

$$I, \rho \Vdash ST_x(\varphi).$$

Hence, by Proposition 9.10, $RT(I), \rho(x) \Vdash \varphi$ and so φ is satisfiable in modal logic K. Hence ST_x is a reduction from

$$\text{Given } \varphi \in L_{ML}^P, \text{ is } \varphi \text{ satisfiable in K?}$$

to

$$\text{Given } \psi \in L_{\Sigma_{ML}^P}, \text{ is } \psi \text{ satisfiable in } \{x, y\}\text{-FOL?}$$

where $\{x, y\}$-FOL is the fragment of first-order logic over x and y and without function symbols. The thesis follows by Proposition 2.14 since ST_x is a computable map and satisfiability in $\{x, y\}$-FOL is decidable (see [22]). $\qquad\qquad\square$

Chapter 10

Kolmogorov Complexity

Kolmogorov in [36] proposed to measure the complexity of an object by the length of the simplest description of the object. For the purpose of this chapter the objects are functions and the descriptions are programs that compute them. The objective is to obtain upper bounds for the minimum length of a program that computes a given function over W. Such programs are enumerated by the indices of oracular universal functions. Recall Section 6.1 for computation with oracles.

10.1 Requirements

We start by showing that our computational model and the universe W fulfil certain requirements that should be present when discussing Kolmogorov complexity (see [38, 46]).

Proposition 10.1. (R0) The universe W is a decidable set.

Proof. The reader is referred to Exercise 2.1. □

Proposition 10.2. (R1) There is a computable embedding $\lambda w . \overline{w} : W \to W$ with decidable range and such that

$$\begin{cases} \exists c \in \mathbb{N} \ \forall w \in W \quad |\overline{w}| \leq |w| + c \\ \forall w, w' \in W \quad |w| < |w'| \text{ iff } |\overline{w}| < |\overline{w'}| \end{cases}.$$

Proof. Take $\overline{w} = "w"$ and $c = 2$. Then $\lambda w . "w"$ is computable and for every $w \in W$, $|"w"| \leq |w| + 2$. The second assertion follows immediately. □

Proposition 10.3. (R2) There is a computable embedding $\lambda k . \hat{k} : \mathbb{N} \to W$ such that:

$$\begin{cases} \forall k \in \mathbb{N} & |\hat{k}| \leq \log_2(k+1)+1 \\ \forall j,k \in \mathbb{N} & j < k \text{ iff } \hat{j} < \hat{k} \end{cases}.$$

Proof. Take $\hat{k}$ as the decimal representation of k. The two statements follow straight-forwardly.   □

Proposition 10.4. (R3) There is a computable embedding

$$\tau : \bigcup_{n \in \mathbb{N}} \mathsf{W}^n \to \mathsf{W},$$

with computable projection

$$\mathsf{Proj} = \lambda\, z, w \begin{cases} w_i & \text{if } \begin{cases} z = \hat{i} \\ w = \tau(w_1,\ldots,w_n) \\ 1 \leq i \leq n \end{cases} \\ \text{undefined} & \text{otherwise} \end{cases} : \mathsf{W}^2 \rightharpoonup \mathsf{W},$$

such that, for each $n \in \mathbb{N}$, $\tau|_{\mathsf{W}^n} : \mathsf{W}^n \to \mathsf{W}$ satisfies

$$\exists c \in \mathbb{N}\ \forall w_1,\ldots,w_n \in \mathsf{W}\ |\tau(w_1,\ldots,w_n)| \leq |w_1| + \cdots + |w_n| + c.$$

Proof. (1) For each $n \in \mathbb{N}$, let

$$\tau|_{\mathsf{W}^n}(w_1,\ldots,w_n) = \langle w_1,\ldots,w_n \rangle.$$

Each $\tau|_{\mathsf{W}^n}$ is an injective map for every n and τ is also an injective map.

(2) It is immediate to see that τ is a computable map.

(3) The projection is a computable function since τ is computable.

(4) Let $n \in \mathbb{N}$. Take $c = n+1$. It is immediate that $|\tau(w_1,\ldots,w_n)| \leq |w_1| + \cdots + |w_n| + c$ since c includes the $n-1$ commas plus the two angle brackets.   □

For the sequel we need to define a family where each element is an oracular universal function $\mathrm{univ}_{\mathsf{W}}^F$ relative to a finite set of oracles F. First we introduce some useful notation.

Remark 10.1. In this chapter, by $\mathscr{F}_1$, $\mathscr{C}_1$, $\mathscr{C}_2$ and $\mathrm{m}\mathscr{C}_1$ we mean the classes of unary functions, unary computable functions, binary computable functions and unary computable maps over W, respectively.

Definition 10.1. Let $F \subseteq \mathscr{F}_1$ be a finite set of oracles. Consider the function:

$$\mathrm{univ}_{\mathsf{W}}^F : \mathsf{W} \times \mathsf{W} \rightharpoonup \mathsf{W}$$

computed relatively to F by the following program $\mathsf{P}_{\mathrm{univ}_W^F}$

$$\text{function } (p, w) \ (\text{return } p(w))$$

So

$$\mathrm{univ}_W^F = \lambda\, p, w\,. \begin{cases} w' & \text{if } w' \text{ results from the execution of } p \text{ on } w \\ \text{undefined} & \text{otherwise} \end{cases}.$$

Proposition 10.5. Let $F \subseteq \mathscr{F}_1$ be a finite set. Then, univ_W^F is universal with respect to $\mathscr{C}_1/F$.

We omit the proof of this result since it follows by a straightforward adaptation of the proof of Proposition 3.2. Observe that univ_W^F enjoys the *s-m-n/F* property. That is, for every $v \in \mathscr{C}_2/F$, there exists $s : W \to W \in \mathrm{m}\mathscr{C}_1$ such that

$$v_p = (\mathrm{univ}_W^F)_{s(p)}$$

for every $p \in W$. Thus univ_W^F is a proper universal function relatively to F.

Remark 10.2. Given a finite set $F \subseteq \mathscr{F}_1$ and $p \in W$, we denote by

$$\phi(p/F)$$

the function $(\mathrm{univ}_W^F)_p$. Furthermore, we denote by $\phi(p)$ the function $(\mathrm{univ}_W^\emptyset)_p$.

10.2 Kolmogorov Complexity of Functions

In this section, we start by defining Kolmogorov complexity of a function with respect to a set of oracles and then prove several results concerning bounds namely for operations between functions.

Definition 10.2. Let $F \subseteq \mathscr{F}_1$ be a finite set and $g \in \mathscr{C}_1/F$. Then,

$$\mathbf{D}(g/F) = \min\{|p| : p \in W, \phi(p/F) = g\}$$

is said to be the *Kolmogorov complexity of g relatively to F*. Moreover,

$$\mathbf{D}(g) = \mathbf{D}(g/\emptyset)$$

is said to be the *Kolmogorov complexity of g*.

Example 10.1. Recall the programs P and P' introduced in Example 1.16 for computing function $f = \lambda x. \frac{x}{2} : \mathbb{N} \to \mathbb{N}$. Then

$$\mathbf{D}(f) \leq |\mathsf{P}| = 68 \quad \text{and} \quad \mathbf{D}(f) \leq |\mathsf{P}'| = 59.$$

Proposition 10.6. Let $F \subseteq \mathscr{F}_1$ be a finite set and $p \in W$. Then,

$$\mathbf{D}(\phi(p/F)/F) \le |p|.$$

Proof. Assume that $\phi(p/F) = g$. Since p computes g relatively to F we can conclude that $\mathbf{D}(\phi(p/F)/F) \le |p|$. $\square$

Remark 10.3. We denote by

$$\pi(g/F)$$

a program that computes g relatively to F with length $\mathbf{D}(g/F)$ (a *witness* for $\mathbf{D}(g/F)$).

We start by characterizing the Kolmogorov complexity of constant unary computable functions.

Proposition 10.7. Let $F \subseteq \mathscr{F}_1$ be a finite set. Then,

(R4) $\exists c \in \mathbb{N} \; \forall y \in W \; \exists p \in W$

$$\begin{cases} \phi(p/F) = \lambda z.y \\ |p| \le |y| + c \end{cases} \quad ;$$

(DR4) $\exists c \in \mathbb{N} \; \forall y \in W \; \mathbf{D}(\lambda z.y/F) \le |y| + c.$

Proof. (R4) Let p be function (z) (return y). Take $c = |\text{function } (z) \text{ (return)}|$. Therefore, $|p| = |y| + c$ and $\phi(p/F) = \lambda z.y$.

(DR4) Let c and p be as in (R4). Then, $\mathbf{D}(\lambda z.y/F) \le |p|$, by Proposition 10.6. Hence $\mathbf{D}(\lambda z.y/F) \le |y| + c$. $\square$

Requirement (R4) tells us that the size of a program computing a constant map $\lambda z.y$ is less than or equal to the length of y up to a constant c that does not depend on y. Thus, (DR4) states that the Kolmogorov complexity of $\lambda z.y$ relatively to F is the length of y modulo a constant c that does not depend on y.

Proposition 10.8. Let $F \subseteq \mathscr{F}_1$ be a finite set. Then,

(R5) $\exists c \in \mathbb{N} \; \forall p \in W \; \forall w \in W \; \exists q \in W$

$$\begin{cases} \phi(q/F) = \phi(\phi(p)(w)/F) \\ |q| \le |p| + |w| + c \end{cases} \quad ;$$

(DR5) $\exists c \in \mathbb{N} \; \forall s \in \mathrm{m}\mathscr{C}_1 \; \forall w \in W \; \mathbf{D}(\phi(s(w)/F)/F) \le \mathbf{D}(s) + |w| + c.$

Proof. (R5) Let q be the program

$$p(w)$$

and $c = |()|$. Therefore, it is immediate to see that $|q| = |p| + |w| + c$ and $\phi(q/F) = \phi(\phi(p)(w)/F)$.

(DR5) Let $s \in m\mathscr{C}_1$. Then, by (R5), exists $q \in W$ such that

$$\phi(q/F) = \phi(s(w)/F) \quad \text{and} \quad |q| \le |\pi(s)| + |w| + c.$$

Therefore, $\mathbf{D}(\phi(s(w)/F)/F) \le \mathbf{D}(s) + |w| + c$, since $\mathbf{D}(\phi(s(w)/F)/F) \le |q|$, by Proposition 10.6, and since $|\pi(s)|$ is $\mathbf{D}(s)$. $\qquad\square$

Requirement (R5) provides a bound on the Kolmogorov complexity of using the *s-m-n* property.

Proposition 10.9. Let $F, G \subseteq \mathscr{F}_1$ be finite sets such that $F \subseteq G$. Then:

(R6) if $F \subseteq G$ then $\forall\, p \in W\ \exists\, q \in W$

$$\begin{cases} \phi(q/G) = \phi(p/F) \\ |q| \le |p| \end{cases} \quad ;$$

(DR6) $\forall\, h \in \mathscr{C}_1/F\ \mathbf{D}(h/G) \le \mathbf{D}(h/F)$;

(DR6w) $\forall\, y \in W\ \mathbf{D}(\lambda z.y/\lambda z.x) \le \mathbf{D}(\lambda z.y)$, where $x \in W$.

Proof. (R6) Assume that $F \subseteq G$ and let $p \in W$. It is enough to take q equal to p.
(DR6) Let $h \in \mathscr{C}_1/F$. Then, by (R6) with $p = \pi(h/F)$ we get

$$\phi(q/G) = \phi(\pi(h/F)/F) \text{ and } |q| \le |\pi(h/F)|.$$

Since $\mathbf{D}(h/G) = |\pi(h/G)| \le |q|$, by Proposition 10.6, the thesis follows.
(DR6w) Let $F = \emptyset$ and $G = \{\lambda z.x\}$. Then, by (DR6),

$$\mathbf{D}(\lambda z.y/G) \le \mathbf{D}(\lambda z.y/F).$$

Hence, $\mathbf{D}(\lambda z.y/\lambda z.x) \le \mathbf{D}(\lambda z.y/\emptyset)$. $\qquad\square$

Requirement (R6) states that a function computable with respect to a finite set of oracles F is always computable with respect to a a finite set $G \supseteq F$ of oracles.

Proposition 10.10. Let $F \subseteq \mathscr{F}_1$ be a finite set. Then

(R7) $\exists\, c \in \mathbb{N}\ \forall\, q, p' \in W\ \exists\, p \in W$;

$$\begin{cases} \phi(p/F) = \phi(q/\phi(p'/F), F) \\ |p| \le |q| + |p'| + c \end{cases} \quad ;$$

(DR7) $\exists c \in \mathbb{N}\ \forall f \in \mathscr{C}_1/F\ \forall g \in \mathscr{C}_1/f, F\ \mathbf{D}(g/F) \leq \mathbf{D}(g/f, F) + \mathbf{D}(f/F) + c;$

(DR7w) $\exists c \in \mathbb{N}\ \forall x, y \in W\ \mathbf{D}(\lambda z.y) \leq \mathbf{D}(\lambda z.y/\lambda z.x) + \mathbf{D}(\lambda z.x) + c.$

Proof. (R7) Let $q, p' \in W$. Let p be the program

$$\mathsf{subs}(q, \mathsf{Oracle}(\phi(p'/F)), p').$$

Take c as

$$|\mathsf{subs}(, \mathsf{Oracle}(\phi(p'/F)),)|.$$

Then $|p| \leq |q| + |p'| + c$ and $\phi(p/F) = \phi(q/\phi(p'/F), F)$.

(DR7) Take $q = \pi(g/f, F)$ and $p' = \pi(f/F)$ in (R7). Then, there is $p \in W$ such that

$$|p| \leq |\pi(g/f, F)| + |\pi(f/F)| + c \text{ and } \phi(p/F) = \phi(q/\phi(p'/F), F).$$

The thesis follows by Proposition 10.6.

(DR7w) Let $F = \emptyset$. Then, let $c \in \mathbb{N}$ be such that

$$\forall f \in \mathscr{C}_1\ \forall g \in \mathscr{C}_1/f\quad \mathbf{D}(g) \leq \mathbf{D}(g/f) + \mathbf{D}(f) + c.$$

Let $x, y \in W$. Observe that $(\lambda z.x), (\lambda z.y) \in \mathrm{m}\mathscr{C}_1$, and so $(\lambda z.y) \in \mathscr{C}_1/f$. The result follows from (DR7) by replacing g and f by $(\lambda z.y)$ and $(\lambda z.x)$, respectively. $\quad\square$

Requirement (R7) states that if a function is computable relatively to F and a computable oracle $\phi(p'/F)$ then it is computable relatively to F.

Proposition 10.11. Let $F \subseteq \mathscr{F}_1$ be a finite set. Then:

(R8) $\exists c \in \mathbb{N}\ \forall p \in W\ \exists q \in W$

$$\begin{cases} \phi(q/h, F) = \phi(p/F) \circ h\quad \forall h \in \mathscr{F}_1 \\ |q| \leq |p| + c \end{cases}.$$

Proof. (R8) Let $p \in W$. Take q to be the program

$$\mathsf{function}\ (z)\ (\mathsf{return}\ p(\mathsf{Oracle}(h)(z)))$$

and $c = |\mathsf{function}\ (z)\ (\mathsf{return}\ (\mathsf{Oracle}(h)(z)))|$. Then, the thesis follows. $\quad\square$

Requirement (R8) tells us that if f is computable with respect to F then by giving h as an oracle, together with F we can compute $f \circ h$.

Proposition 10.12. Let $F \subseteq \mathscr{F}_1$ be a finite set. Then:

(R9) $\exists c \in \mathbb{N} \ \forall q \in W \ \exists p \in W$

$$\begin{cases} \phi(p/F) \circ \lambda z.\langle z,x \rangle = \phi(q/\lambda z.x, F) & \forall x \in W \\ |p| \le |q| + c \end{cases} .$$

Proof. Let $q \in W$. Take p to be the program

function (w) (return subs$(q, \text{Oracle}(\lambda z. w[2]), \text{function}(z)(\text{return } w[2]))(w[1]))$

and

$c = |\text{function}(w)(\text{return subs}(, \text{Oracle}(\lambda z. w[2]), \text{function}(z)(\text{return } w[2]))(w[1]))|.$

Thus, the thesis follows. $\qquad\qquad\qquad\qquad\qquad\qquad\qquad\qquad\qquad \square$

Clearly, (R9) is a very restricted form of oracle elimination.

10.3 Kolmogorov Complexity of Operations

We now investigate the relationship between the Kolmogorov complexity of composition and aggregation with the Kolmogorov complexity of the component functions.

Proposition 10.13. Let $F \subseteq \mathscr{F}_1$ be a finite set. Then:

(wPo) $\exists c \in \mathbb{N} \ \forall p_1, p_2 \in W \ \exists q \in W$

$$\begin{cases} \phi(q/F) = \phi(p_2/F) \circ \phi(p_1/F) \\ |q| \le |p_1| + |p_2| + c \end{cases} ;$$

(wDPo) $\exists c \in \mathbb{N} \ \forall f,g \in \mathscr{C}_1/F \quad \mathbf{D}(g \circ f/F) \le \mathbf{D}(g/F) + \mathbf{D}(f/F) + c.$

Proof. (wPo) Choose

$$v = \lambda p,x. (\phi(\text{Proj}(2,p)/F) \circ \phi(\text{Proj}(1,p)/F))(x) : W \times W \to W.$$

Observe that

$$v(\tau(p_1, p_2), x) = (\phi(p_2/F) \circ \phi(p_1/F))(x).$$

It is straightforward to see that v is computable relatively to F. Since univ_W^F is a proper universal function relatively to F, there is $s \in \mathrm{m}\mathscr{C}_1$ such that

$$v(p,x) = \phi(s(p)/F)(x)$$

for every $p \in W$ and $x \in W$. Hence, in particular,

$$v(\tau(p_1, p_2), x) = \phi(s(\tau(p_1, p_2))/F)(x)$$

for every $p_1, p_2 \in W$ and $x \in W$. Thus,

$$\phi(p_2/F) \circ \phi(p_1/F) = \phi(s(\tau(p_1, p_2))/F)$$

for every $p_1, p_2 \in W$. Choose p_0 such that

$$\phi(p_0) = s.$$

Then,

$$\phi(p_2/F) \circ \phi(p_1/F) = \phi(\phi(p_0)(\tau(p_1, p_2))/F)$$

for every $p_1, p_2 \in W$. On the other hand, by (R5), there exists $c' \in \mathbb{N}$ such that for every $p_1, p_2 \in W$ there is $q \in W$ such that

$$\begin{cases} \phi(q/F) = \phi(\phi(p_0)(\tau(p_1, p_2))/F) \\ |q| \leq |p_0| + |\tau(p_1, p_2)| + c' \end{cases}.$$

Thus, thanks to (R3), there are $c', c'' \in \mathbb{N}$ such that for every $p_1, p_2 \in W$ there is $q \in W$ such that:

$$\begin{cases} \phi(q/F) = \phi(\phi(p_0/F)(\tau(p_1, p_2))/F) \\ |q| \leq |p_0| + |p_1| + |p_2| + c'' + c' \end{cases}.$$

The thesis follows by taking $c = |p_0| + c' + c''$.

(wDP∘) Let $f, g \in \mathscr{C}_1/F$. Then, by (wP∘) with $\pi(f/F)$ and $\pi(g/F)$, there are $c \in \mathbb{N}$ and $q \in W$ such that

$$\phi(q/F) = \phi(\pi(g/F)/F) \circ \phi(\pi(f/F)/F) \quad \text{and} \quad |q| \leq |\pi(f/F)| + |\pi(g/F)| + c.$$

Thus, $|q| \leq \mathbf{D}(f/F) + \mathbf{D}(g/F) + c$ and by Proposition 10.6, the thesis follows.   □

Proposition 10.14. For any finite set $F \subseteq \mathscr{F}_1$:

(sP∘) $\exists c \in \mathbb{N} \ \forall p_1, p_2 \in W \ \exists q \in W$

$$\begin{cases} \phi(q/F) = \phi(p_2/\phi(p_1/F), F) \circ \phi(p_1/F) \\ |q| \leq |p_1| + |p_2| + c \end{cases}.$$

Proof. For arbitrary $p_1, p_2 \in W$, let

$$\begin{cases} f = \phi(p_1/F) \\ g = \phi(p_2/\phi(p_1/F), F) = \phi(p_2/f, F) \end{cases}.$$

By (R8), choosing p such that $\phi(p/F) = id_W$, there is q_0 such that

$$h = \phi(q_0/h, F)$$

for every $h \in \mathscr{F}_1$. Hence, in particular, we have:

$$f = \phi(q_0/f, F).$$

By (wP∘), there is $c' \in \mathbb{N}$ such that for any $p_2 \in W$ there is $q' \in W$ such that:

$$\begin{cases} \phi(q'/f, F) = \phi(p_2/f, F) \circ \phi(q_0/f, F) = g \circ f \\ |q'| \le |q_0| + |p_2| + c' \end{cases}.$$

By (R7), there are $c'' \in \mathbb{N}$ such that for any $p_1 \in W$ there is $q \in W$ such that:

$$\begin{cases} \phi(q/F) = \phi(q'/\phi(p_1/F), F) = \phi(q'/f, F) = g \circ f \\ |q| \le |q'| + |p_1| + c'' \end{cases}.$$

Hence, there are $c', c'' \in \mathbb{N}$ such that for any $p_1, p_2 \in W$ there is $q \in W$ such that:

$$\begin{cases} \phi(q/F) = \phi(q'/f, F) = g \circ f \\ |q| \le |q_0| + |p_2| + c' + |p_1| + c'' \end{cases}.$$

The thesis follows by taking $c = |q_0| + c' + c''$. □

Exercise 10.1. For any finite set $F \subseteq \mathscr{F}_1$, show that:

(sDP∘) $\exists c \in \mathbb{N} \; \forall f, g \in \mathscr{C}_1/F \; \mathbf{D}(g \circ f / F) \le \mathbf{D}(g/f, F) + \mathbf{D}(f/F) + c$

(sDP∘') $\exists c \in \mathbb{N} \; \forall f, g \in \mathscr{C}_1/F \; \mathbf{D}(g \circ f / F) \le \mathbf{D}(g/F) + \mathbf{D}(f/g, F) + c.$

We leave to the interested reader the proof of the following statements related to the Kolmogorov complexity of aggregation.

Exercise 10.2. Let $F \subseteq \mathscr{F}_1$ be a finite set. Show that

(wP⟨⟩) $\exists c \in \mathbb{N} \; \forall p_1, p_2 \in W \; \exists q \in W$

$$\begin{cases} \phi(q/F) = \langle \phi(p_1/F), \phi(p_2/F) \rangle \\ |q| \le |p_1| + |p_2| + c \end{cases} ;$$

(wDP⟨⟩) $\exists c \in \mathbb{N} \; \forall f, g \in \mathscr{C}_1/F \; \mathbf{D}(\langle f, g \rangle / F) \le \mathbf{D}(f/F) + \mathbf{D}(g/F) + c.$

Exercise 10.3. Let $F \subseteq \mathscr{F}_1$ be a finite set. Show that

(sP⟨⟩) $\exists c \in \mathbb{N} \; \forall p_1, p_2 \in W \; \exists q \in W$

$$\begin{cases} \phi(q/F) = \langle \phi(p_1/F), \phi(p_2/\phi(p_1/F), F) \rangle \\ |q| \le |p_1| + |p_2| + c \end{cases} ;$$

(sDP⟨⟩) $\exists c \in \mathbb{N} \; \forall f, g \in \mathscr{C}_1/F \; \mathbf{D}(\langle f, g \rangle / F) \le \mathbf{D}(f/F) + \mathbf{D}(g/f, F) + c;$

(sDP⟨⟩') $\exists c \in \mathbb{N} \; \forall f, g \in \mathscr{C}_1/F \; \mathbf{D}(\langle f, g \rangle / F) \le \mathbf{D}(f/g, F) + \mathbf{D}(g/F) + c.$

10.4 Kolmogorov Complexity of Sets

Since a set is listable if and only if its characteristic function is computable, it is easy to extend our notion of Kolmogorov complexity to listable sets relatively to a finite set of oracles. This notion extends the one in Definition 7.8.

Definition 10.3. Let $F \subseteq \mathscr{F}_1$ be a finite set. We say that $D \subseteq \mathscr{S}_1$ is a *listable set relatively to* F whenever χ_D^p is a computable function relatively to F.

Definition 10.4. Let $F \subseteq \mathscr{F}_1$ be a finite set. We say that the *Kolmogorov complexity of D relatively to* F is

$$\mathbf{D}(D/F) = \mathbf{D}(\chi_D^p/F) = \min\{|p| : \phi(p/F) = \chi_D^p\}.$$

Proposition 10.15. Let $F \subseteq \mathscr{F}_1$ be a finite set. Then,

$$\exists c \in \mathbb{N} \ \forall X, Y \in \mathscr{L}_1/F \ \mathbf{D}(X \cap Y/F) \leq \mathbf{D}(X/F) + \mathbf{D}(Y/F) + c.$$

Proof. Observe that

$$\chi_{X \cap Y}^p = \times \circ \langle \chi_X^p, \chi_Y^p \rangle.$$

Thus, taking into account (wDP$\circ$), see Proposition 10.13, and (wDP$\langle\rangle$), see Exercise 10.2, there are $c', c'' \in \mathbb{N}$ such that for every $X, Y \in \mathscr{L}_1/F$ we have:

$$\begin{aligned}
\mathbf{D}(X \cap Y/F) &= \mathbf{D}(\chi_{X \cap Y}^p/F) \\
&= \mathbf{D}(\times \circ \langle \chi_X^p, \chi_Y^p \rangle/F) \\
&\leq \mathbf{D}(\times/F) + \mathbf{D}(\langle \chi_X^p, \chi_Y^p \rangle/F) + c' \\
&\leq \mathbf{D}(\times/F) + \mathbf{D}(\chi_X^p/F) + \mathbf{D}(\chi_Y^p/F) + c' + c''.
\end{aligned}$$

The thesis follows by taking $c = \mathbf{D}(\times/F) + c' + c''$. $\square$

Since the complement of a decidable set is also decidable and, so, listable, one should examine the relationship between the Kolmogorov complexities of a decidable set and its complement.

Proposition 10.16. $\exists c \in \mathbb{N} \ \forall Y \in \mathscr{D}_1 \ \mathbf{D}(\chi_Y) \leq \mathbf{D}(Y) + \mathbf{D}(Y^c) + c.$

Proof. We start by proving that for any set Y we have

$$\chi_Y \in \mathscr{C}_1/\{\chi_Y^p, \chi_{Y^c}^p\}.$$

Indeed the following program P:

```
function (w) (
    k = 1;
    r = true;
    s = true;
    while r ∧ s do (
        if stceval(Oracle(χᵖ_Y)(w),k,−1) ≠ −1 then r = false;
        if stceval(Oracle(χᵖ_{Yᶜ})(w),k,−1) ≠ −1 then s = false;
        k = k+1
    );
    if r then return 1 else return 0
)
```

computes χ_Y relatively to χ_Y^p and $\chi_{Y^c}^p$. Then,

$$\mathbf{D}(\chi_Y/\chi_Y^p,\chi_{Y^c}^p) \leq |\mathsf{P}|.$$

Assume that Y is a decidable set. Then, χ_Y is a computable map. Hence, by (DR7), there are $c',c'' \in \mathbb{N}$ such that

$$\mathbf{D}(\chi_Y) \leq \mathbf{D}(\chi_Y/\chi_Y^p) + \mathbf{D}(\chi_Y^p) + c'$$

and

$$\mathbf{D}(\chi_Y/\chi_Y^p) \leq \mathbf{D}(\chi_Y/\chi_Y^p,\chi_{Y^c}^p) + \mathbf{D}(\chi_{Y^c}^p/\chi_Y^p) + c''.$$

Thus, by (DR6),

$$\mathbf{D}(\chi_Y) \leq \mathbf{D}(\chi_Y^p) + \mathbf{D}(\chi_{Y^c}^p) + |\mathsf{P}| + c' + c''.$$

Therefore, taking $c = |\mathsf{P}| + c' + c''$ the result follows. $\square$

Exercise 10.4. For any $F \subseteq \mathscr{F}_1$, show that:

$$\forall c \in \mathbb{N} \; \exists Z \in \mathscr{D}_1 \quad \mathbf{D}(Z/F) + c < \mathbf{D}(Z^c/F).$$

Bibliography

[1] S. Aanderaa and D. E. Cohen. Modular machines, the word problem for finitely presented groups and Collins' theorem. In *Word Problems, II*, volume 95 of *Studies in Logic and the Foundations of Mathematics*, pages 1–16. North-Holland, 1980.

[2] S. Arora and B. Barak. *Computational Complexity*. Cambridge University Press, 2009.

[3] A. S. Asratian, T. M. J. Denley, and R. Häggkvist. *Bipartite Graphs and Their Applications*. Cambridge University Press, 1998.

[4] H. P. Barendregt. *The Lambda Calculus: Its Syntax and Semantics*. Studies in Logic and the Foundations of Mathematics. North-Holland, 1984.

[5] P. Blackburn and J. F. A. K. van Benthem. Modal logic: A semantic perspective. In P. Blackburn, J. F. A. K. van Benthem, and F. Wolter, editors, *Handbook of Modal Logic*, pages 173–204. Elsevier Science, 2006.

[6] B. Bollobas. *Modern Graph Theory*. Graduate Texts in Mathematics. Springer, 1998.

[7] G. Boolos, J. Burgess, and R. C. Jeffrey. *Computability and Logic*. Cambridge University Press, third edition, 1999.

[8] D. S. Bridges. *Computability*, volume 146 of *Graduate Texts in Mathematics*. Springer-Verlag, 1994.

[9] A. Church. An unsolvable problem of elementary number theory. *American Journal of Mathematics*, 58(2):345–363, 1936.

[10] A. Church. An unsolvable problem of elementary number theory. *Journal of Symbolic Logic*, 1(2):73–74, 1936.

[11] A. Church. *The Calculi of Lambda-Conversion*. Annals of Mathematics Studies. Princeton University Press, 1941.

[12] D. E. Cohen. *Computability and Logic*. John Wiley, 1987.

[13] S. Barry Cooper. *Computability Theory*. Chapman and Hall, 2003.

[14] B. J. Copeland, C. J. Posy, and O. Shagrir. *Computability: Turing, Gödel, Church, and Beyond*. MIT Press, 2013.

[15] R. Cori and D. Lascar. *Mathematical Logic, Part II, Propositional Calculus, Boolean Algebras, Predicate Calculus and Completeness Theorems*. Oxford University Press, 2000.

[16] H. S. M. Coxeter. *Introduction to Geometry*. John Wiley & Sons, second edition, 1969.

[17] N. J. Cutland. *Computability: An Introduction to Recursive Function Theory*. Cambridge University Press, 1980.

[18] M. Davis. *Computability and Unsolvability*. Dover, 1958.

[19] R. L. Epstein and W. A. Carnielli. *Computability: Computable Functions, Logic and the Foundations of Mathematics*. Wadsworth, second edition, 2000.

[20] K. Gödel. *Collected Works: Vol. I*. The Clarendon Press, Oxford University Press, 1986. Publications 1929-1936.

[21] K. Gödel. *Collected Works. Vol. II*. The Clarendon Press, Oxford University Press, 1990. Publications 1938-1974.

[22] E. Grädel, P. G. Kolaitis, and M. Y. Vardi. On the decision problem for two-variable first-order logic. *The Bulletin of Symbolic Logic*, 3(1):53–69, 1997.

[23] E. R. Griffor, editor. *Handbook of Computability Theory*, volume 140 of *Studies in Logic and the Foundations of Mathematics*. North-Holland, 1999.

[24] H. N. Gupta. *Contributions to the Axiomatic Foundations of Geometry*. ProQuest LLC, 1965. Thesis (Ph.D.)–University of California, Berkeley.

[25] R. Herken, editor. *The Universal Turing Machine: A Half-Century Survey*. The Clarendon Press, Oxford University Press, 1988.

[26] H. Hermes. *Enumerability Computability Decidability*. Springer, 1965.

[27] D. Hilbert. *Grundlagen der Geometrie*, volume 6 of *Teubner-Archiv zur Mathematik*. B. G. Teubner Verlagsgesellschaft mbH, fourteenth edition, 1999.

[28] D. Hilbert and W. Ackermann. *Grundzüge der theoretischen Logik*. Springer-Verlag, fifth edition, 1972.

[29] D. Hilbert and W. Ackermann. *Principles of Mathematical Logic*. American Mathematical Society, 2003.

[30] G. A. Jones and J. M. Jones. *Elementary Number Theory*. Springer Undergraduate Mathematics Series. Springer-Verlag, 1998.

[31] J. P. Jones and Y. V. Matijasevič. Register machine proof of the theorem on exponential Diophantine representation of enumerable sets. *The Journal of Symbolic Logic*, 49(3):818–829, 1984.

[32] J. L. Kelley. *General Topology*. Springer-Verlag, 1975.

[33] S. C. Kleene. General recursive functions of natural numbers. *Mathematische Annalen*, 112(1):727–742, 1936.

[34] S. C. Kleene. Recursive predicates and quantifiers. *Transactions of the American Mathematical Society*, 53:41–73, 1943.

[35] S. C. Kleene. *Mathematical Logic*. John Wiley & Sons, 1967.

[36] A. N. Kolmogorov. On tables of random numbers. *SankhyĀ*, 25:369–376, 1963.

[37] S. Lang. *Algebra*, volume 211 of *Graduate Texts in Mathematics*. Springer-Verlag, third edition, 2002.

[38] M. Li and P. Vitányi. *An Introduction to Kolmogorov Complexity and Its Application*. Springer, third edition, 2008.

[39] S. Mac Lane. *Categories for the Working Mathematician*, volume 5 of *Graduate Texts in Mathematics*. Springer-Verlag, second edition, 1998.

[40] D. Marker. *Model Theory: An Introduction*, volume 217 of *Graduate Texts in Mathematics*. Springer-Verlag, 2002.

[41] Y. V. Matijasevič. Enumerable sets are Diophantine. *Doklady Akademii Nauk SSSR*, 191:279–282, 1970.

[42] Y. V. Matiyasevich. *Hilbert's Tenth Problem*. MIT Press, 1993.

[43] E. Mendelson. *Introduction to Mathematical Logic*. Chapman and Hall, sixth edition, 2015.

[44] M. L. Minsky. Size and structure of universal Turing machines using tag systems. In *Proceedings of the Sympososium on Pure Mathematics, V*, pages 229–238. American Mathematical Society, 1962.

[45] M. L. Minsky. *Computation: Finite and Infinite Machines*. Prentice-Hall, 1967.

[46] A. Nies. *Computability and Randomness*, volume 51 of *Oxford Logic Guides*. Oxford University Press, 2009.

[47] A. Olszewski, J. Wolenski, and R. Janusz, editors. *Church's Thesis After 70 Years*, volume 1 of *Ontos Mathematical Logic*. Ontos Verlag, 2007.

[48] E. L. Post. Finite combinatory processes-formulation 1. *Journal of Symbolic Logic*, 1(3):103–105, 1936.

[49] E. L. Post. Formal reductions of the general combinatorial decision problem. *American Journal of Mathematics*, 65:197–215, 1943.

[50] B. Robič. *The Foundations of Computability Theory*. Springer, 2015.

[51] J. Robinson. *The Collected Works*. American Mathematical Society, 1996.

[52] H. Rogers, Jr. *Theory of Recursive Functions and Effective Computability*. MIT Press, second edition, 1987.

[53] Y. Rogozhin. Small universal Turing machines. *Theoretical Computer Science*, 168(2):215–240, 1996.

[54] W. Rudin. *Principles of Mathematical Analysis*. McGraw-Hill, third edition, 1976.

[55] A. Sernadas and C. Sernadas. *Foundations of Logic and Theory of Computation*. College Publications, London, second edition, 2012.

[56] A. Shen and N. K. Vereshchagin. *Computable Functions*. American Mathematical Society, 2003.

[57] J. C. Shepherdson and H. E. Sturgis. Computability of recursive functions. *Journal of the Association for Computing Machinery*, 10:217–255, 1963.

[58] J. R. Shoenfield. *Mathematical Logic*. Association for Symbolic Logic, 2001.

[59] R. I. Soare. *Turing Computability: Theory and Applications*. Springer, 2016.

[60] A. Syropoulos. *Hypercomputation: Computing Beyond the Church-Turing Barrier*. Springer, 2008.

[61] A. Tarski. *A Decision Method for Elementary Algebra and Geometry*. University of California Press, second edition, 1951.

[62] A. Tarski and S. Givant. Tarski's system of geometry. *Bulletin of Symbolic Logic*, 5(2):175–214, 1999.

[63] A. Turing. On computable numbers, with an application to the Entschei-dungsproblem. *Proceedings of the London Mathematical Society*, 2(42):230–265, 1936.

[64] A. Turing. *Pure Mathematics*. North-Holland, 1992.

List of symbols

Subject index